# 建筑业信息分解编码
## A&bCode

黄 强 著
A&bCode 研究组

中国建筑工业出版社

图书在版编目（CIP）数据

建筑业信息分解编码A&bCode：汉英对照 / 黄强，A&bCode研究组著. — 北京：中国建筑工业出版社，2019.4

ISBN 978-7-112-23354-0

Ⅰ.①建… Ⅱ.①黄…②A… Ⅲ.①建设设计—计算机辅助设计—应用软件—汉、英 Ⅳ.① TU201.4

中国版本图书馆CIP数据核字（2019）第034188号

本书用中英文双语编写，图文并茂，内容包括BIM的互操作性特征、建筑信息分类编码（OmniClass）、建筑业信息分解编码（A&bCode）、基于A&bCode的HIM实现互操作性、A&bCode与BIM等。本书可供工程技术人员、软件开发者、建筑类高等院校师生及BIM相关人员阅读参考。

责任编辑：武晓涛
责任校对：姜小莲

## 建筑业信息分解编码A&bCode

黄　强
A&bCode研究组　著

\*

中国建筑工业出版社出版、发行（北京海淀三里河路9号）
各地新华书店、建筑书店经销
北京点击世代文化传媒有限公司制版
天津图文方嘉印刷有限公司印刷

\*

开本：787×1092毫米　1/16　印张：12¼　插页：2　字数：260千字
2019年4月第一版　2019年4月第一次印刷
定价：130.00元
ISBN 978-7-112-23354-0
（33663）

**版权所有　翻印必究**
如有印装质量问题，可寄本社退换
（邮政编码 100037）

## A&bCode 研究组成员

黄　强　柏文杰　吴露方　张　良　杨嵩桥　杨龙龙
高子斌　赵　锋　王照华　王　荣　张　淼　都　浩
左　睿　柳　跃　叶方甦　沈丽丽

## A&bCode Research Goup Members

HUANG Qiang, BAI Wenjie, WU Lufang,
ZHANG Liang, YANG Songqiao, YANG Longlong,
GAO Zibin, ZHAO Feng, WANG Zhaohua,
WANG Rong, ZHANG Miao, DU Hao,
ZUO Rui, LIU Yue, YE Fangsu, SHEN Lili

# 序

《道德经》有云:"大方无隅,大象无形",这是老子"道"的至高境界。世界上万事万物往往都不拘泥于自己的基因,表现出"气象万千"的形态与格局,信息技术也是如此,看似晦涩,想来深奥,但结合工作实际却是简单。这是普遍存在的、辨证的唯物主义世界观。

不同软件间的互操作(功能对接,传递数据,模型互通,信息协作)可狭义定义为:一个软件将数据送到另外一个软件。从广义来看则为:不同软件之间可以互相调取并有效使用对方的数据。但无论如何,在不同软件间联通的对象只能是数据,即使是某个有形的三维构件形体,在软件中也是以数据的形式表达。当我们把电脑(硬件)关闭,软件会把所有的三维模型折好,叠放到相应空间里面去,即将数据保存到某个数据库中。

因此,互操作性(Interoperability)即是软件间的数据互用。

美国BIM标准体系的目标是软件的互操作性(互用性)。

美国工业互联网参考架构(IIRA)指出:工业互联网将致力于工业控制系统联网,使之形成大型的端对端系统。将其与人联系起来,能充分集成企业内部系统、业务流程和分析解决方案。这些端对端系统被称为工业互联网系统(IISs)。在工业互联网系统内部,操作传感器数据和人员与系统的互动可与组织或公共信息结合起来,用于高级分析和其他高级数据处理(例如基于规则的政策和决策系统)。这种分析和处理的结果反过来又将使大量日益自主的控制系统在优化决策、操作和协作方面取得重大进展。工业互联网系统由许多从不同厂商和组织提供的部分所组成。要使得这些部分成功组合在一起,它们必须具备以下性质:

集成性——指的是部件通过各种信号和协议与其他部件相连的能力。

互用性——基于通用概念构架和统一情境内的信息理解能力。

兼容性——按照重组要求与其他组件互动,并达到互动方要求的能力。

显然,兼容性(Compatibility)是工业互联网的最优实现方式。兼容性有赖于并有助于互用性和集成性。

中国P-BIM标准体系的目标是软件的兼容性。

大道至简。在进行系统分析时,重点不仅仅是了解整体,还需深入了解事物内部的结构和组成，各个组成部分之间的集成和协同关系。为了完成这个任务，最重要的工作就是做到互相独立、完全穷尽、层级粒度适当的分解，以达到庖丁解牛，目无全牛而游刃有余。分解的过程是一个自上而下的过程，而分类和抽象的过程恰好是一个自下而上的过程。分解的目的是由整体到个体，同时通过个体的分析来洞悉事物内在运行机制；分类的目的则是从个体到整体，通过分类和抽象来实现对抽象类别的统一决策和行动。因此，不对建筑业信息进行分解的分类必然是个庞大、难以控制的分类。建筑业信息的自上而下分解与适当的层级粒度、自下而上的分类及统一决策的结合，就在于建筑业子行业系统纲要性工作分解结构（SWBS）。

悟在天成。为使同一系统中的不同软件（子系统）满足兼容性要求，要对系统进行纲要性工作分解（任务）并对系统中的每项任务建立不同功能子系统编码（code）体系，利用子系统编码间的逻辑关系就可以实现不同子系统软件间文件夹的双向交互，同时对交换文件夹定义分类交互内容和交换数据格式标准，即可实现软件的兼容性。

1993年ISO/TR 14177题为《建筑业信息分类》的技术报告，开启了建筑业信息分类旅程；时隔26年，我们将创新性开启《建筑业信息分解编码A&bCode》以适应建筑业信息化发展需要。

面向对象方法中，把从具有共同性质的实体中抽象出的事物本质特征概念，称为类（Class）；面向系统工程方法中，把系统中的成员视为具有自身目的与主动性的、积极的"活的"主体，称为码（Code）。通过A&bCode重构建筑业软件体系功能，使建筑业软件工具模块化，做小、做简、做精建筑业专业软件；利用A&bCode建立建筑业信息网络交换系统，朝着"网络架构极简、网络交易模式极简、网络极安全、隐私保护遵从GDPR"这四个目标要求，做大、做好建筑业互联网网络系统。

类似于互联网网络架构，建立建筑业互联网网络系统架构遵循以下原则：以建筑业网络、设备、系统数据交换接口编码A&bCode对等于互联网公网的固定IP地址；以A&bCode系列P-BIM软件功能与信息交换标准对等于互联网网络传输协议家族TCP/IP；以基于A&bCode的建筑业互联网HIM网络操作系统对等于互联网的域名系统DNS。

## 国内外主要建筑业分类与编码体系发展时间轴

**从构件分类编码到活动分解编码、2018中国建筑业信息分解编码自主技术元年**

**1972年以UniFormat开发为标志开始的国内外建筑业信息分类编码发展历程**

# 目 录

序

## 第1章　BIM 的互操作性特征 ... 001
1.1　基于文本与基于模型的系统工程 ... 002
1.2　对等网络 ... 003
1.3　美国 BIM 标准信息交换架构 ... 004
1.4　从"点对点"到"一对一"再回"点对点" ... 004
1.5　OpenBIM ... 007

## 第2章　建筑信息分类编码（OmniClass） ... 008
2.1　线分法与面分法 ... 008
2.2　MasterFormat 与 UniFormat ... 009
2.3　建筑信息分类编码 OmniClass 与 UniClass ... 011
2.4　分类编码用于 BIM 存在问题 ... 017

## 第3章　建筑业信息分解编码（A&bCode） ... 020
3.1　分解与分类 ... 020
3.2　工作分解结构（Work Breakdown Structure，WBS） ... 024
3.3　模式 ... 028
3.4　建筑业任务分解体系 ... 030
3.5　建筑信息模型分解结构
　　　（模型分解结构，Model Breakdown Structure，MBS） ... 032
3.6　分布式功能建模软件（P-BIM 功能软件） ... 034
3.7　P-BIM 功能软件信息交换标准 ... 037
3.8　建筑业信息分解编码体系（A&bCode） ... 037
3.9　A&bCode 核心思想及其意义 ... 041
3.10　A&bCode 编码标准制定示范（公路工程） ... 042

## 第 4 章　基于 A&bCode 的 HIM 实现互操作性 ·········· 045
　　4.1　数字索网 ·········· 045
　　4.2　基于 A&bCode 的 HIM 数字索网 ·········· 047
　　4.3　HIM 实现兼容性和互操作性 ·········· 050

## 第 5 章　A&bCode 与 BIM ·········· 054
　　5.1　美国 BIM 标准体系与中国 BIM 标准体系 ·········· 054
　　5.2　A&bCode 与 OmniClass ·········· 055
　　5.3　A&bCode 与 IFC、NBIMS 信息交换架构 ·········· 055
　　5.4　A&bCode 与 IDM/MVD ·········· 064

致　　谢 ·········· 069
A&bCode 软件数据交换演示 ·········· 075
英文版 ·········· 077
附　　图：A&bCode 与建筑业互联网 ·········· 插页

# 第 1 章　BIM 的互操作性特征

美国 BIM 标准第一版第一部分指出：

软件的可互操作性是在不同应用软件之间信息的无缝交换，其中每个应用程序中都有它自己的内部数据结构。实现可互操作性是通过将每个参与应用程序的内部数据结构的组成部分映射（mapping）到通用的数据模型，反之亦然。如果采用的通用数据模型是开放性的，任何应用程序都可以参与映射过程，因而与参与映射过程的其他任何应用程序形成了可互操作关系。可互操作性消除了将每个应用程序(不同版本的程序)与别的应用程序(不同版本的程序)的集成工作，大大降低了数据的应用成本。

美国 BIM 标准的参考标准提供了基础的、独立于计算机的那些实体、属性、关系和分类的定义，对表达建筑行业的丰富语言十分关键。NBIMS 委员会选定的参考标准都是国际标准，就共享复杂设计和施工项目内容的能力来讲，已取得理想的效果。NBIMS 第一版第一部分包括三部候选参考标准：国际协同工作联盟（IAI）的工业基础分类标准（IFC）、美国建筑标准协会(CSI)的建筑信息分类体系标准( OmniClass )和美国建筑标准协会( CSI )的国际语义框架标准（IFD）。

IFC 数据模型由定义、规则和协议组成，它们以独特方式定义了描述建筑全生命期的数据集。这些定义允许行业软件开发者将 IFC 接口写入他们的软件，实现与其他软件应用程序交流和分享相同格式的信息，不管其他软件应用程序的内部数据结构如何。有 IFC 接口的软件应用程序之间也能交换和共享信息。

建筑信息分类编码体系标准（OmniClass 或 OCCS）是用于基本建设行业的多表分类系统。OmniClass 包括基本建设行业最常用的分类法，适用于组织不同形式的美国 BIM 标准的重要信息，有电子版的，也有硬拷贝。OCCS 可以用于准备许多类型的项目信息和交换的信息、成本信息、规格信息和在建筑全生命期中所产生的其他信息。

bSDD（bS 数据字典，buildingSMART Data Dictionary）是一种必须用于多种语言环境但能实现取得一致结果的建筑行业术语词典。美国 NBIMS 的设计依赖术语和分类一致（通过 OmniClass）来支持模型的互操作。OmniClass 表中的输入项目一旦在 IFDLibrary 中定义后可多次重复使用，实现软件应用程序之间可靠的自动沟通——这就是美国 BIM 标准的主要目标。

国际 buildingSMART 组织 openBIM，以及美国 BIM 标准的 BIM 共同目标是 Interoperability（互操作性）（图1-1）。

图 1-1　BIM 的目标：Interoperability（互操作性）

## 1.1　基于文本与基于模型的系统工程

传统的系统工程就是以文档为中心的系统工程（图1-2），这个文档又是"基于文本的"，所以也可以说传统的系统工程是"基于文本的系统工程"（Text-based Systems Engineering，TSE）。

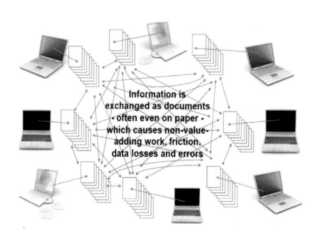

图 1-2　基于文本的系统工程

在《系统工程 2020 年愿景》中，给出了"基于模型的系统工程（图1-3）"的定义：基于（系统架构）模型的系统工程是对建模（活动）的形式化应用（formalized application of modeling），以便支持系统要求的设计、分析、验证和确认等活动，这些活动从概念设计阶段开始，持续贯穿到设计开发以及后来的全生命期阶段。

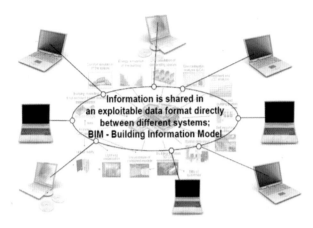

图 1-3　基于（系统架构）模型的系统工程

## 1.2　对等网络

对等网络（P2P），即对等计算机网络，是一种在对等者（Peer）之间分配任务和工作负载的分布式应用架构，是对等计算模型在应用层形成的一种组网或网络形式。

简单地说，P2P 就是直接将使用者联系起来，让使用者通过互联网直接交互。P2P 使得网络上的沟通变得容易、更直接共享和交互，真正地消除中间商。P2P 另一个重要特点是改变互联网现在的以大网站为中心的状态、重返"非中心化"，并把权力交还给用户。

对等网络是一种网络结构的思想，它与目前网络中占据主导地位的客户端或浏览器/服务器（Client/Server 简称 C/S, Browser/Server 简称 B/S）架构（也就是 WWW 所采用的结构方式）的一个本质区别是，整个网络结构中不存在中心节点（或中心服务器）。在 P2P 结构中，每一个节点（peer）大都同时具有信息消费者、信息提供者和信息通信这三方面的功能。从计算模式上来说，P2P 打破了传统的 Client/Server（C/S）或 B/S 模式，在网络中的每个节点的地位都是对等的。每个节点既充当服务器，为其他节点提供服务，同时也享用其他节点提供的服务。

与 C/S 或 B/S 网络相比，对等网络具有下列优势：

1）可在网络的中央及边缘区域共享内容和资源。在 C/S 或 B/S 网络中，通常只能在网络的中央区域共享内容和资源。

2）由对等方组成的网络易于扩展，而且比单台服务器更加可靠。单台服务器会受制于单点故障，或者会在网络使用率偏高时，形成瓶颈。

3）由对等方组成的网络可共享处理器，整合计算资源以执行分布式计算任务，而不只是单纯依赖一台计算机，如一台超级计算机。

4）用户可直接访问对等计算机上的共享资源。网络中的对等方可直接在本地存储器上共享文件，而不必在中央服务器上进行共享。

## 1.3 美国BIM标准信息交换架构

美国BIM标准关于数据交换的层次架构如图1-4所示。信息交换架构分为信息交付模型活动、模型视图、派生视图和聚合视图四个层次，在模型视图层次中，IFC通过MVD定义一项活动或商业论证的信息交换方式，它将所有IFC概念中的类、属性、关系、属性集、数量定义和其他，使其在该子集使用，它代表了软件需求规范。通过实现IFC接口（中间文件格式）以满足交换要求。而各层级间的交换则需要独立的特定IFC版本，实现更大范围（多重）IFC的类、属性、关系、属性集、数量定义和其他的子集、交集、并集、补集的信息集合运算。

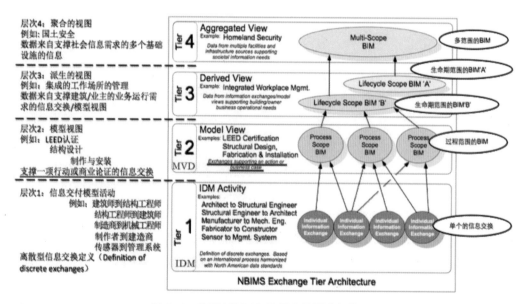

图1-4 美国NBIMS数据交换层次架构图

## 1.4 从"点对点"到"一对一"再回"点对点"

我们经常用图1-5把传统的"点对点"（基于文本）的文本交换方式简单地描述为"一对一"（基于模型）的交换方式：

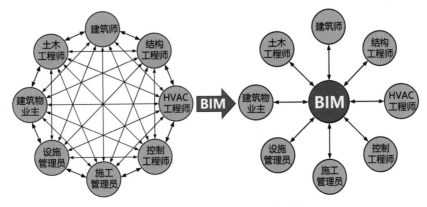

图 1-5 "点对点"与"一对一"

实际上，图 1-5 存在一个严重的误区，左侧是基于文本交换，是人与人之间的交换，而右侧，在 BIM 工作模式下，应当是不同专业人员使用软件才能与 BIM 数据库进行交换，因此，图 1-5 右侧 BIM 对应的对象是应用软件而不是具体的人。

准确来说，图 1-5 右侧 BIM 对应的对象是软件，左侧的文本交换也应该换为软件间的交换才能对应对比，即图 1-5 左侧应该是图 1-6。

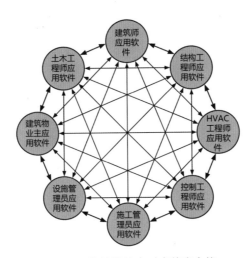

图 1-6 软件间的点对点信息交换

天下没有免费的午餐。工程信息交换的本质不会因为简单地画一个图而改变。从图 1-4 可见，图 1-5 的各专业人员信息交换本质并没有改变，仅仅是集中图 1-5 左侧的文本交换（人类语言，IDM）转化为计算机语言（模型视图，MVD）的过程，这个过程是基于图 1-5 左侧原有的交换内容编制 IDM，再集中某个应用软件的所有 IDM 制定 MVD，按照美国 NBIMS 的数据交换层次架构图（图 1-4），图 1-5 右侧"BIM"落地实施路径应该如图 1-7 所示。

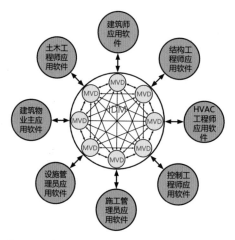

图 1-7　美国 NBIMS BIM 落地实施路径图

综上，我们应该把图 1-5 纠正为图 1-8。

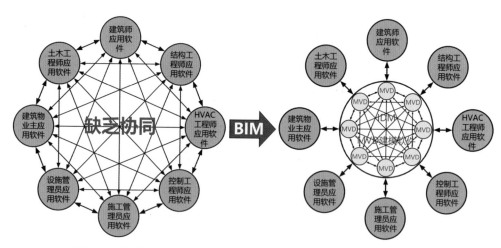

图 1-8　从"点对点"到"一对一"（美国 NBIMS）的软件间数据交换

由以上介绍，可以认识到应用对等网络形式进行不同软件间的数据交换方式将解决现在 BIM 实施方式存在的诸多问题，而图 1-8 左侧的交换方式正好是典型的 P2P 对等网络交换方式。

以图 1-8 左侧的方式实施 BIM 的最大问题是不同软件间的信息交换缺乏协同，由于在建设工程实施过程中的不同阶段，"协同"内容不同。如：在设计阶段的协同主要内容是不同专业间的空间协同，在施工阶段的协同主要是不同工序之间的进度和工作面的协同。以图 1-6 左侧的方式实施 BIM，就需要补充一个广义的"协同软件"（图 1-9）。

图 1-9 右侧还是 P2P 对等网络方式。

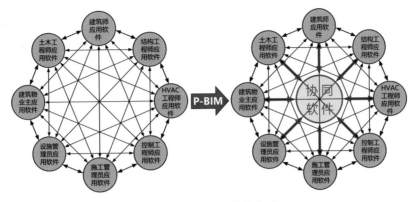

图 1-9　P-BIM 实施方式

## 1.5　OpenBIM

Open BIM 是基于开放标准和工作流进行协同设计、建筑实现和运营的一个普遍方式。在与国际 buildingSMART 组织主席和总裁的交流中谈到 openBIM 的实施方式目标是实现两个软件间的数据无缝对接（如图 1-10 所示），从图 1-10 可以看到 openBIM 的数据交换是基于单个 IDM 直接实现 MVD 交付数据至接收方软件的应用方式，这种做法更像图 1-9 左侧，也是 P2P 方式。

国际 buildingSMART 组织正试图建立一个 MVD 库，供不同软件使用（如图 1-11 所示）。

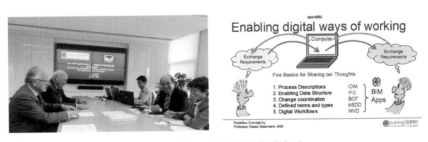

图 1-10　openBIM 的数据流方式

图 1-11　buildingSMART 的 BIM 实现方式设想

# 第 2 章　建筑信息分类编码（OmniClass）

建立编码体系的目的在于对建设项目全过程进行科学有效的管理，规范工程参与者的行为。具体而言，它有利于项目建设单位对项目各个阶段工作内容的控制，如有助于对工程总造价进行管理控制、实行价值工程研究、为项目各成员提供信息交流工具，尤其是为建设单位、设计单位、施工单位之间信息沟通提供一种共同语言，在有效传达信息的同时，消除误解。另外，工程信息分类编码为工程项目数据收集、汇总、整理和分析的基础，为未来项目使用准确的、有价值的信息提供了保证。

## 2.1　线分法与面分法

线分类法，是按选定的若干属性（或特征）将分类对象逐次地分为若干层级，每个层级又分为若干类目。统一分支的同层级类目之间构成并列关系，不同层级类目之间构成隶属关系。同层级类目互不重复，互不交叉。

优点:(1)操作更人性化，符合传统应用习惯，既适合于手工处理，又便于计算机处理;(2)扩容性好;(3)编码的检索非常高效;(4)数据可以分级管理，层次性好，能较好地反映类目之间的逻辑关系。

缺点:(1)揭示主题或事物具体特征的能力差，往往无法满足确切分类的需要，不能充分体现目前大量存在的细小分类问题;(2)分类表具有一定的凝固性，不便于根据需要随时改变，也不适合进行多角度的信息检索;(3)无法根据现代科学的发展自动生成新类，难以与科学的发展保持同步;(4)大型分类表一般类目详尽、篇幅较大，对分类表管理的要求较高;(5)分类结构弹性差。

面分类法也称平行分类法，是把拟分类的对象集合总体，根据其本身固有的属性或特征，分成相互之间没有隶属关系的面，每个面都包含一组类目。将某个面中的一种类目与另一个面的一种类目组合在一起，即组成一个复合类目。

优点: 主要优点是分类结构上具有较大的柔性，即分类体系中任何一个"面"内类目的变动，不会影响其他"面"，而且可以对"面"进行增删。再有，"面"的分类结构可根据任意"面"的组合方式进行检索，这有利于计算机的信息处理。面分类法还具有可以较大量地扩充、结构弹性好、不必预先确定好最后的分组、适用于计算机管理等优点。

缺点：主要缺点是组配结构太复杂、不便于手工处理，不能充分利用编码空间。

## 2.2 MasterFormat 与 UniFormat

1963 年美国 CSI 与加拿大 CSC，共同发布了 CSI 建筑规范格式，目标是用于项目档案管理、造价管理和编写组织规范。最早的 MasterFormat 版本一共有 16 个分类，于 1978 年发布第一版，这个版本主要是面向房建项目，一直维护到 1995 年。之后，随着很多新的产品、材料和工艺涌入建筑业，机械和管道越来越复杂，同时也有越来越多的市政和工业项目需要编码。这些新的信息都无法编排到 MasterFormat 1995 里面。于是在 2004 年，CSI 发布了 MasterFormat 2004 版，把原来的 16 个类别扩展为 50 个类别，又把这 50 个类归纳到六个分组里（表 2-1）。

MasterFormat 分类表　　　　　　　　　　　表 2-1

| | |
|---|---|
| 招投标与合同需求组 | |
| 00 招投标与合同需求 | |
| 通用需求 | |
| 01 通用需求 | |
| 设施建设组 | |
| 02 现场条件 | 03 混凝土工程 |
| 04 砌体工程 | 05 金属工程 |
| 06 木材、塑料和复合材料工程 | 07 保温防水工程 |
| 08 门窗工程 | 09 装饰工程 |
| 10 建筑配件 | 11 设备工程 |
| 12 室内用品 | 13 特殊施工 |
| 14 运输系统 | |
| 设施服务 | |
| 21 消防设施 | 22 管 道 |
| 23 暖通和空调 | 25 综合自动化 |
| 26 电 气 | 27 通 信 |
| 28 电子安全和保安 | |
| 场地和基础设施组 | |
| 31 土方工程 | 32 外部改造 |
| 33 市政工程 | 34 运输工程 |
| 35 港口和航道工程 | |

续表

| 工艺设施组 | |
|---|---|
| 40 过程互连 | 41 材料加工处理设备 |
| 42 加热、冷却、干燥设备 | 43 气体、液体处理、净化和存储设备 |
| 44 污染和废物控制设备 | 45 特殊行业制造设备 |
| 46 水和废水处理设备 | 48 发电设备 |

MasterFormat 是一种面向材料和工种的建筑信息分类方法，建设单位使用它进行任务分解、成本计算、招标投标的时候可便捷使用。但到了投资方和设计方的手里，材料和工序这些小类目，在方案估算、限额设计、动态成本控制几个方面，则会造成障碍。所以，美国建筑师学会（AIA）和美国总务管理局（GSA）按照建筑物组成元素的分类思路各自开发的两套编码体系，后来把各自的标准整合到一起，于 1972 年共同开发，命名为 UniFormat。现行的 UniFormat 有两个版本，一个是 ASTM 发布的 UniFormat II，于 1993 年首次发布，最新版是 2015；另一个是 CSI 和 CSC 发布的 UniFormat（表 2-2），最新版是 2010。这两个都是从 AIA 和 GSA 联合发布的最初版 UniFormat 发展而来的。MasterFormat 与 UniFormat 都采用了线分法。

**CSI 版 UniFormat 第一层：主要元素组**　　　　表 2-2a

| | | |
|---|---|---|
| A | 基础结构 | |
| B | 外封闭工程 | |
| C | 建筑内部 | |
| D | 配套设施 | |
| E | 设备及家具 | |
| F | 特殊建筑和建筑拆除 | |
| G | 建筑场地 | |
| Z | 一般要求（CSI 独有） | |

**CSI 版 UniFormat2010 前三层分解**　　　　表 2-2b

| Level1 | Level2 | Level3 |
|---|---|---|
| A 基础结构 | A10 基础 | A1010 一般基础 |
| | | A1020 特殊基础 |
| | A20 地下围护结构 | A2010 地下围护墙 |
| | A40 底板 | A4010 标准底板 |
| | | A4020 结构底板 |
| | | A4030 底板基槽 |

续表

| Level1 | Level2 | Level3 |
|---|---|---|
| A 基础结构 | A40 底板 | A4040 底板基坑 |
| | | A4090 底板附属构件 |
| | A60 地下室排水工程 | A6010 建筑地下排水 |
| | | A6020 地下排气 |
| | A90 基础结构相关活动 | A9010 基础结构开挖 |
| | | A9020 施工排水 |
| | | A9030 基坑支护 |
| | | A9040 土壤治理 |

## 2.3 建筑信息分类编码 OmniClass 与 UniClass

### 2.3.1 Omni Class

1993 年，题为《建筑业信息分类》ISO/TR 14177 的技术报告中指出，原有编码体系的分类范围不能涵盖建筑业的各个方面，是不完整的体系。这份报告提出了基于面分类法的建筑信息分类体系框架，定义了一些新的建筑分类对象，比如设施、空间、设计构件、工项、产品、辅助工具、建设活动等，为现代建筑编码体系奠定了基础。实际上提出了针对建筑对象分类的两种方式：线分类法和面分类法。要描述像建筑业这样复杂的对象，仅仅依靠有前后层次所属关系的线分类法是不够的，还有更多的分类方式并不存在前后关联性，需要通过面分类法来进行描述，类似于在拓扑学里有向图和无向图关系。

1996 年，以 ISO/TR 14177 以及实际的工程经验为基础，ISO 组织发布了一份重要的标准：ISO 12006-2。提出了一个基本过程模型，将建筑分为"建设过程"、"建设资源"和"建设成果"三个大分类。这实际上是用三个维度对建筑业全生命期进行数字化描述的一种尝试（如图 2-1 所示）。

美国建筑规范协会 CSI 基于 ISO 12006-2 标准，参考建筑业分类编码相关的继承资源，牵头建立了北美建筑业信息模型分类编码体系——OmniClass。这份由建筑业相关的 17 个组织共同起草的编码标准从 2000 年开始制定，历经 6 年才发布了 1.0 版。

OmniClass 采用面分法与线分法相结合的方式，共有 15 张分类表。这 15 张表（如表 2-3 所示）是以面分类法进行分类。只是从不同角度对建筑进行认识，这几个部分有一定的关联性，但是没有从属关系。每张分类表内部采用线分法，代表着一种建筑信息的分类方法。

这种分类方法一方面能够适应建筑信息复杂多样的特点，另一方面又能够充分继承已有的各种传统分类的成果。由于面分类法各自独立，相互不关联，所以采用独立文件的方式进行保存。

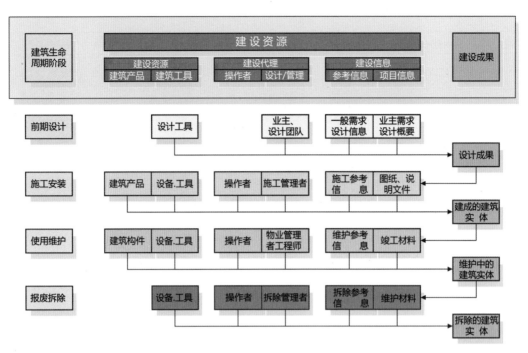

图 2-1 ISO 12006-2 对建筑业全生命期的数字化描述

OmniClass 的 15 张表格　　　　　　　　　　　　　　　　表 2-3

| 表格 11- 按功能定义的建筑实体 | 表格 32- 服务 |
| --- | --- |
| 表格 12- 按形式定义的建筑实体 | 表格 33- 学科 |
| 表格 13- 按功能定义的空间 | 表格 34- 组织角色 |
| 表格 14- 按形式定义的空间 | 表格 35- 工具 |
| 表格 21- 元素 | 表格 36- 信息 |
| 表格 22- 工作成果 | 表格 41- 材料 |
| 表格 23- 产品 | 表格 49- 属性 |
| 表格 31- 阶段 | |

　　OmniClass 虽然由 15 个表格组成，最常用的表格 23- 产品表是 Revit 软件中构件的分类编码基础，在 Revit 软件安装目录下可提取出 "OmniClass Taxonomy" 文件（如图 2-2 所示），打开之后可以看到其构件的分类编码以表代码 23 起始（如图 2-3 所示）。但也仅有表格 23 得到 Autodesk 的原生嵌入，究其原因，无论是继承 UniFormat 的表格 21- 元素，继承 MasterFormat 的表格 22- 工作成果，亦或是继承 EPIC 的表格 23- 产品，OmniClass 分类编码体系始终围绕"构件"之间的逻辑关系而展开，缺少对策划、规划、设计、合约、实施（竣工）、运维各阶段完整"建物流"要素的集成，面分法产生的 15 个表格间信息断层依旧存在。另一方面，在不同的面分表格间，通过逻辑运算符将编码联合

来描述复杂建筑对象时，如描述"实施阶段的儿童图书馆的板柱框架结构柱"，需要表格 31+ 表格 11+ 表格 23 的编码组合"31-60 00 00+11-12 29 14+23-13 35 11 13 11"，如此冗长的 28 位编码仅用来表示"建筑工程全生命期范围内某一个特定对象"，将导致计算机存储效率、信息传递效率的下降。

图 2-2　OmniClassTaxonomy　　　　图 2-3　Revit 内置构件编码

OmniClass 是 MasterFormat 和 UniFormat 是两种分类方法的融合，其结果如图 2-4 所示。

图 2-4　OmniClass 的基本组合

OmniClass 标准依然缺乏一种清晰地将 15 张表紧密统一成为整体的思想。

美国 BIM 标准委员会也于 2016 年指出由于 NBIMS 缺乏一套清晰的层次框架结构，无法形成统一的建设工程全生命期业务流程，导致标准未得到软件商的实施，如图 2-5 所示。

图 2-5 NBIMS 面临的挑战

### 2.3.2 UniClass

由英国 CPIC 开发的 UniClass 是面分法与线分法结合的建筑统一分类编码，UniClass 2015 已发布的表格组成如表 2-4 所示。目前由 NBS 负责更新和维护，其分类编码内容融汇于 OmniClass 15 个表格当中，即 UniClass 与 OmniClass 互为继承资源，OCCS 开发委员会可以实际需求引用并改编 UniClass 的内容，同时这种交叉也使得 UniClass 可以 OmniClass 为参考实现进一步的优化。

UniClass 是一套动态的建筑业统一分类编码标准（更新至 2018 年 10 月 28 日） 表 2-4

| Table | 状态和修订信息 |
| --- | --- |
| Co-Complexes（综合体） | 2018 年 8 月发布 V1.7 |
| En-Entities（实体） | 2018 年 10 月发布 V1.10 |
| Ac-Activities（活动） | 2018 年 10 月发布 V1.8 |
| SL-Spaces/locations（空间/位置） | 2018 年 10 月发布 V1.10 |
| EF-Elements/functions（元素/功能） | 2018 年 8 月发布 V1.3 |
| Ss-Systems（系统） | 2018 年 10 月发布 V1.12 |
| Pr-Products（产品） | 2018 年 10 月发布 V1.12 |
| TE-Tools and Equipment（工具和设备） | 2018 年 8 月发布 V1.5 |
| PM-Project management（项目管理） | 2018 年 8 月发布 V1.2 |
| Zz-CAD（CAD） | 2015 年 7 月发布 V1.0 |
| FI-Form of information（信息形式） | 测试当中 |

无论是 CPIC 的 UniClass 或 CSI/CSC 的 OmniClass，由于分类编码成果的不完整性，无法均匀地服务于建筑工程的所有行业类别，也无法满足建筑工程全生命期范围的信息集成要求。相较于 OmniClass，NBS 负责更新和维护的 UniClass 2015 具有许多提升，

（1）OmniClass 表格以 PDF 和 Excel 格式发布，相较于纸质编码，一定程度上便于检索与使用，而 UniClass 2015 以数字格式呈现，允许所有表格被同时快速地检索和同义词搜索；（2）OmniClass 的 15 个表格间并不协调统一，多数表格为各方独立开发后汇集整理，而 UniClass 2015 则是一套统一的建筑业分类编码体系，其不同表格间术语、排序、分组结构等保持一致；（3）编码结构方面，OmniClass 表代码由无实际意义的数字组成（如 21,22），代码级别数量从两级至八级，而 UniClass 2015 表代码由具有意义的单词缩写组成（如活动 Ac，实体 En），且代码级别范围大都相对一致地维持在三级至五级（如表 2-5、表 2-6 所示）。

UniClass Table Activities　　　　　　　　　　表 2-5

| Code | Group | Sub group | Section | Object | Title |
|---|---|---|---|---|---|
| Ac_05 | 05 | | | | Project management activities |
| Ac_05_00 | 05 | 00 | | | Strategy stage activities |
| Ac_05_00_10 | 05 | 00 | 10 | | Business case development |
| Ac_05_00_80 | 05 | 00 | 80 | | Strategic brief preparation |
| Ac_05_00_82 | 05 | 00 | 82 | | Strategic brief submission |
| Ac_05_10 | 05 | 10 | | | Brief stage activities |
| Ac_05_10_15 | 05 | 10 | 15 | | Cost estimate preparation |
| Ac_05_10_17 | 05 | 10 | 17 | | Cost estimate submission |
| Ac_05_10_29 | 05 | 10 | 29 | | Feasibility study preparation |
| Ac_05_10_31 | 05 | 10 | 31 | | Feasibility study submission |
| Ac_05_10_61 | 05 | 10 | 61 | | preliminary design preparation |
| Ac_05_10_63 | 05 | 10 | 63 | | preliminary design submission |
| Ac_05_10_65 | 05 | 10 | 65 | | Project brief and objectives preparation |
| Ac_05_10_67 | 05 | 10 | 67 | | Project brief and objectives submission |
| Ac_05_20 | 05 | 20 | | | Concept stage activities |
| Ac_05_20_15 | 05 | 20 | 15 | | Concept cost report preparation |
| Ac_05_20_17 | 05 | 20 | 17 | | Concept cost report submission |
| Ac_05_20_21 | 05 | 20 | 21 | | Concept design development |
| Ac_05_20_23 | 05 | 20 | 23 | | Concept design report preparation |
| Ac_05_20_25 | 05 | 20 | 25 | | Concept design report submission |
| Ac_05_30 | 05 | 30 | | | Definition stage activities |
| Ac_05_30_03 | 05 | 30 | 03 | | Agreement negotiating |
| Ac_05_30_10 | 05 | 30 | 10 | | Building regulation assessing |
| Ac_05_30_21 | 05 | 30 | 21 | | Definition design cost report preparation |
| Ac_05_30_23 | 05 | 30 | 23 | | Definition design cost report submission |

续表

| Code | Group | Sub group | Section | Object | Title |
|---|---|---|---|---|---|
| Ac_05_30_25 | 05 | 30 | 25 | | Definition design development |
| Ac_05_30_27 | 05 | 30 | 27 | | Definition design report preparation |
| Ac_05_30_29 | 05 | 30 | 29 | | Drainage adoption agreeing |
| Ac_05_30_37 | 05 | 30 | 37 | | Highways adoption agreeing |
| Ac_05_30_60 | 05 | 30 | 60 | | Party wall notices agreeing |
| Ac_05_30_64 | 05 | 30 | 64 | | Planning preparation |
| Ac_05_30_85 | 05 | 30 | 85 | | Sustainability assessing |
| Ac_05_40 | 05 | 40 | | | Design stage activities |
| Ac_05_40_85 | 05 | 40 | 85 | | Technical design cost report preparation |
| Ac_05_40_87 | 05 | 40 | 87 | | Technical design development |
| Ac_05_40_89 | 05 | 40 | 89 | | Technical design report preparation |
| Ac_05_50 | 05 | 50 | | | Build and commission stage activities |
| Ac_05_50_15 | 05 | 50 | 15 | | Contractor mobilizing |

UniClass Table Entities  表 2-6

| Code | Group | Sub group | Section | Object | Title |
|---|---|---|---|---|---|
| En_20 | 20 | | | | Administrative, commercial and protective service |
| En_20_10 | 20 | 10 | | | Legislative entities |
| En_20_10_45 | 20 | 10 | 45 | | Governmental buildings |
| En_20_15 | 20 | 15 | | | Administrative office entities |
| En_20_15_10 | 20 | 15 | 10 | | Multiple occupation office buildings |
| En_20_15_70 | 20 | 15 | 70 | | Single occupation office buildings |
| En_20_20 | 20 | 20 | | | Secular representative entities |
| En_20_20_10 | 20 | 20 | 10 | | Buildings for representatives of nation states abroad |
| En_20_20_40 | 20 | 20 | 40 | | Local government buildings |
| En_20_20_50 | 20 | 20 | 50 | | National government buildings |
| En_20_20_70 | 20 | 20 | 70 | | Regional government buildings |
| En_20_45 | 20 | 45 | | | Motor vehicle maintenance and fuelling entities |
| En_20_45_50 | 20 | 45 | 50 | | Motor vehicle fuelling and charging entities |
| En_20_45_54 | 20 | 45 | 54 | | Motor vehicle servicing and repair entities |
| En_20_50 | 20 | 50 | | | Commercial entities |
| En_20_50_05 | 20 | 50 | 05 | | Auction buildings |
| En_20_50_22 | 20 | 50 | 22 | | Department stores |
| En_20_50_29 | 20 | 50 | 29 | | Financial and professional services buildings |
| En_20_50_50 | 20 | 50 | 50 | | Markets |

续表

| Code | Group | Sub group | Section | Object | Title |
|---|---|---|---|---|---|
| En_20_50_53 | 20 | 50 | 53 | | Mixed use buildings |
| En_20_50_55 | 20 | 50 | 55 | | Motor vehicle sales entites |
| En_20_50_80 | 20 | 50 | 80 | | Shop units |
| En_20_50_85 | 20 | 50 | 85 | | Supermarkets |
| En_20_50_97 | 20 | 50 | 97 | | Wholesale buildings |
| En_20_55 | 20 | 55 | | | Postal communications entities |
| En_20_55_65 | 20 | 55 | 65 | | Post office buildings |
| En_20_55_80 | 20 | 55 | 80 | | Sorting office buildings |
| En_20_60 | 20 | 60 | | | Military entities |
| En_20_60_02 | 20 | 60 | 02 | | Air force buildings |
| En_20_60_10 | 20 | 60 | 10 | | Army buildings |
| En_20_60_56 | 20 | 60 | 56 | | Navy buildings |
| En_20_65 | 20 | 65 | | | Law enforcement operational entities |
| En_20_65_09 | 20 | 65 | 09 | | Police buildings |
| En_20_70 | 20 | 70 | | | Judicial entities |
| En_20_70_40 | 20 | 70 | 40 | | Law court buildings |
| En_20_75 | 20 | 75 | | | Detention entities |
| En_20_75_10 | 20 | 75 | 10 | | Detention buildings |
| En_20_80 | 20 | 80 | | | Weapons training ranges |
| En_20_80_29 | 20 | 80 | 29 | | Firing range buildings |
| En_20_80_30 | 20 | 80 | 30 | | Exterior firing ranges |
| En_20_85 | 20 | 85 | | | Security entites |

总的来说，UniClass 2015 的项目阶段和项目管理表体现了建筑工程的全生命期时间线，所有阶段地位等同。且 UniClass 与 ISO 12006-2：2015 系统保持一致，可实现到 NRM1（NRM1——费用估算和分项费用计划规则；NRM2——工程采购的工程量清单规则；NRM3——运营和维护费用计划和采购规则）以及后续的其他分类编码系统的映射。

## 2.4 分类编码用于 BIM 存在问题

MasterFormat 的 1998 年第一版和 UniFormat II 的 1993 年第一版，目标是用于项目档案管理、造价管理和编写组织规范、方案估算、限额设计、动态成本控制几个方面，是各自独立工作需要。而美国 BIM 标准第一版是在 2007 年颁布，显然 OmniClass 不是为 BIM 而生，简单地将 OmniClass 应用于 BIM 是无法取得成功的。

无论是美国 OmniClass 还是英国 UniClass 信息分类方式用于 BIM，必然要有软件商的支持，而编码不全、不完善或不好用等原因都可能让软件商弃之不用而失去编码的意义。

由于建筑工程的复杂性，不可能建立单一完全覆盖建筑物全生命期的应用系统，这个应用系统应该是由上百款不同软件组成，每一款工程应用软件都只是基于特定目的，支持特定阶段的业务工作。而要以 BIM 为中心数据模型，实现不同工程软件的数据交互必须依赖统一的数据标准。在建筑业，目前 IFC（Industry Foundation Classes，工业基础类）架构是最为全面的面向对象的数据模型，涵盖了工程设计领域各个阶段满足全部商业需求的数据定义。IFC 标准的第一个版本于 1997 年 1 月由 IAI 组织（Industry Alliance for Interoperability，现为 BuildingSMART International）发布。然而，在实际的应用中，基于 IFC 的信息分享工具需要能够安全可靠地交互数据信息，但 IFC 标准并未定义不同的项目阶段，不同的项目角色和软件之间特定的信息需求，兼容 IFC 的软件解决方案的执行因缺乏特定的信息需求定义而遭遇瓶颈，软件系统无法保证交互数据的完整性与协调性。针对这个问题的一个 BIM 解决方案，就是制定一套标准，将实际的工作流程和所需交互的信息定义清晰，而这个标准就是 IDM 标准（Information Delivery Manual，信息交付手册）。显然，这个 IDM 是针对"完全覆盖建筑物全生命期的应用系统中实际工作流程单一软件所需的交互信息"，它的目标在于使得针对全生命期某一特定阶段的信息需求标准化，并将需求提供给软件商，与公开的数据标准（IFC）映射，最终形成解决方案。2007 年发布的美国 BIM 标准第一版及 openBIM 方法均以 IFC+IFD+IDM 为实现 BIM 的基本标准。

在这种以 IFC 为标准的中间数据格式实际实施 BIM 过程中，以构件为对象建模并在不同软件间传递信息，形成了以 IFC 标准为主要交互标准的"构件思维"。这种"构件思维"一直主导着 BIM 理论，然而，至今二十年，IFC 的"构件思维"理论并未在实际工程中得到真正应用（图 2-6）。

一个完整项目从策划到竣工有大量复用频率很高的信息，如：设计若无特殊需求时，只用在设计信息中明确某建筑顶层结构柱子混凝土强度等级为 C30，而在随后的施工图审查过程中，混凝土的供应商和质检人员就已经有多套不同的评价规范来对该建筑顶层结构柱子的混凝土实现丰富的信息填充，在造价、质量、浇筑工艺、养护时间、强度评价、检查要点等有多类信息；建设方、设计方、施工方、监理方、试验检验方、监督方等也会以多方视角来规范地描述这个"C30"。面对如此成熟的信息网络，今天的建筑业建造过程还在使用着传统的文本传递信息，这就要求我们重新思考是否有另外一种更好的方式来实现建筑业信息化。

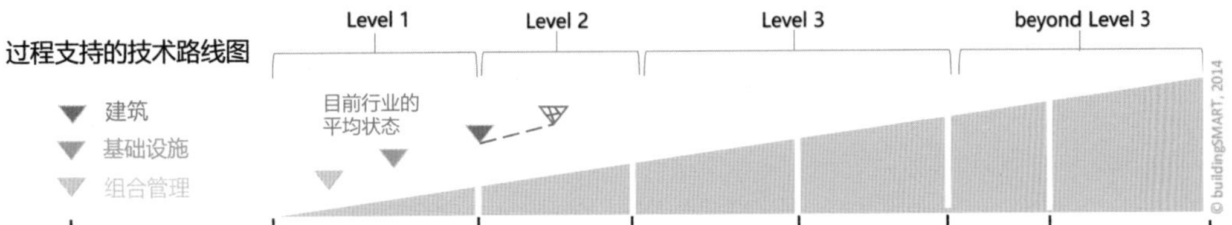

| 过程支持的技术路线图 | Level 1 | Level 2 | Level 3 | | beyond Level 3 | |
|---|---|---|---|---|---|---|
| 主题 | 文件 | 大BIM | 目的BIM和数据提取 | 工作流BIM | 云BIM | … |
| 工作方式 | 2D/3D图纸 | 3D BIM 具体学科文件夹 | | | BIM数据 经网络扩散 | |
| 标准、开放格式 | dxf、dwg、pdf | ifc（CV、COBie） | ifc's mvdXML | ifc's BCF | | |
| 交流方式 | 基于文件工作 | 大模型交换 | 目的驱动的模型交换 | 工作流驱动的模型升级 | | |
| 技术手段 | | 文件夹服务区，引用整个模型 | BIM中心，引用分项模型 | 网络服务，引用对象 | | |
| 任务清单 | | 启用IFC，定义LOD（IDM5+） | 目的MVD（25+）交付首个IFC基础设施 | 模块化IFC，网络链接，交接要求 | | |
| 未来发展 | | | 增强运管阶段 | 准备组合管理 | | |

图 2-6 过程支持的技术路线图

# 第 3 章　建筑业信息分解编码（A&bCode）

ISO/TR 14177 题为《建筑业信息分类》的技术报告指出：建筑业工作要素多，这些要素间的关系呈现相互交错的复杂状态。无论涉及何种组织模式，每个建设过程的总和——策划阶段、实施阶段、使用阶段、运维阶段、拆除阶段都是相同的。每项过程可以被分解为必须执行的活动以推动项目的进行。在分类分析中该过程是主要的：an Action and business（A&b）。对建筑业信息分解时，由于不同阶段、不同专业、不同工具对于数据的要求是不一样的。且不同阶段交付的成果精度要求不同；不同专业交付的成果内容不同；不同工具交付的成果形式不同。因此，简单地以某一种分类标准为基础（例如以构件为基础或者以材料为基础），不可能满足所有相关专业工作的数据交换需求。

若要不同的人都基于项目全生命期展开工作，进行各种形式的数据交换，相互配合，形成协同，最终建成整个建筑业信息系统，就需要有一套完整合理的编码系统（A&bCode），将不同阶段、不同专业、不同工作方法的工作成果，交换文件名及其交换内容与格式标准统一成为一个有机的整体。

## 3.1　分解与分类

分：永远是在并级和层级两个方面进行分割。

分解（分/解）：自上而下，原来是一个整体，这个整体的各个部分能够协同发挥各部分（个体）所不具有的整体功能。把它分成为各部分（个体）以及各部分（个体）之间的相互关系。这个可以分解为各个体与个体之间关系所组成的整体，能发挥各个体单独运行不能发挥的作用。

对于分解有两个重点，即组成单元和组成单元间的集成和协同。

分类（分/类）：自底向上，原来是一个整体，但是构成这个整体的元素更多是空间、时间或特性上的相互接近，其整体特性并不一定大于各元素特性之和并发挥整体作用。把整体分成各个元素，但是元素之间的关系是简单的空间、时间或特性关系。

对于分类同样也有两个重点，其一是分类，其二是抽象，两者缺一不可。

### 3.1.1 分解

分解是把一个整体事物分成各个组成部分（个体）。

在进行单个事物分析的时候，重点不仅仅是了解整体，在了解完整体后还需深入到事物内部，了解事物内部的结构和组成，各个组成部分（个体）之间的集成和协同关系。为了完成这个工作，最重要的工作就是分解。

对于分解，如果讲简单点，和人体解剖往往是一个道理，即从一个外部整体通过各种手段深入到事物内部的组成单元，同时进一步搞清楚这些组成单元之间是如何协同运作的，即事物最终表现出来的各种不同的外部表现都涉及内部单元之间是如何协同运作的。

人运动后会出汗，这是一个表象，但是只有通过深入事物内部，我们才能够明白人体内部各个器官之间是如何协同运作最终导致了这么一个结果出来。这样，你就能从事物表象转变到事物内在机理的认识。

庖丁解牛，目无全牛而游刃有余，则已经是从事物外在深入到了事物内在的运行机理。

要认识到，形成完整的事物分解结构或分解树只是完成了分解的第一步，更加重要的是对于分解完成的各个子件或组成单元（个体），我们还需根据事物的外在表现特征将其串联起来，只有这样你才会发现事物分解后的各个部分并不是孤立的，而是紧密地协同在一起的。

对于分解而言，往往存在两种顺序问题：

1）静态到动态：首先进行分解，再去研究分解后各个组成单元（个体）间的协同；

2）动态到静态：先观察事物的动态运行过程，找到各个具体的单元组件，再考虑事物如何科学分解。

这两种分解方式，以对于类似人体结构研究而言，我们很容易就会想到直接进行解剖方式分解就可以了；但是对一个新的知识领域或问题，你有时候连究竟应该分解为哪些知识点或问题点都不清楚，这个时候首先要研究问题的动态形成过程，从动态分析的过程中找到分解单元。

但是不论哪种顺序的分解，都必须解决分解为个体，个体间如何协同两大问题。

在事物分解的时候，还有一个重要概念，就是对于复杂的事物，不论是实际事物还是抽象的知识，都可能存在一种多维结构，当存在这种多维结构的时候，我们一定要学会将多维结构转换为多个二维结构，因为只有二维结构我们更加容易进行可视化的分析，这本身就是一个将事物从复杂到简单的一个转化过程。

要把一个事物描述清楚，在现实生活里面我们最容易理解的还是平面化的结构化图形或者形象图形，而对于立体模型本身理解起来相对更加困难。我们可以看下立体几何和平面几何，立体几何的学习困难度大于平面几何，同时对于立体问题我们首先是期望将其转化为平面问题，难点在于立体如何转到平面。大学理工科专业往往会开设机械制图课，对于一个立体结构，我们比较容易输出 3 个视角的投影图，但如果给你 3 个视角的平面图，你能否快速地构想出具体的立体形象？即很多时候对于事物的分析，分解容易而集成难，而一个完整的闭环应该是：

立体多维→二维或单维分析→各个维度集成分析→聚合还原并闭环。

这个思路清楚后可以看到对于工作中很多场景都基于该分析思路。拿一个软件系统来说，当我们对其架构进行分析和设计的时候，可以看到架构本身就存在多个维度和视角，业务流程、数据、技术、物理部署、运行机制等都可能是架构关键的视角。要把架构描述清楚，就必须首先对这些单个维度逐一地进行分析和描述。但是给出一个完整的架构，其核心不在于单独的每个架构视图描述清楚，而是在于各个架构视图之间是如何集成的，各个架构视图如何能够整合为一个完整的架构。架构的难点已经不在单架构视图，而在于对单视图的整合能力，只有理清了如何整合，才能更好地从静态分析到动态分析，拿架构来说即静态模型到架构内部的动态运行机制。

任何静态事物的分解在分解的时候都必须考虑层级的概念，不同的层级下应该具备相同的粒度，在分解的时候最容易犯错的两个地方，其一是不符合 MECE（Mutually Exclusive Collectively Exhaustive，相互独立，完全穷尽）法则，MECE 法则是对于一个重大的议题，能够做到不重叠、不遗漏的分解，而且能够借此有效把握问题的核心，并解决问题的方法；其二就是分解层级粒度问题。这两个问题都解决好了才是一个完整的结构分解，这种分解才能够更好地根据金字塔原理的思路去做最终的呈现。要明白最终呈现本身的不清楚核心原因在于分解时候的结构不清楚，演绎本身的模糊也在于最终归纳上的错误。

静态事物的分解和呈现最终都是无法真正理清事物的内在机制，因此对事物本身的动态和运行机制分析往往才是更加重要的内容。对事物的动态分析一个是把事物看为一个整体，根据时间线去研究事物本身的发展演进过程。当你做了这一步后，会衍生出第二个问题，即是什么推动了事物的这种动态变化？任何事物的动态发展变化一方面是外在环境和事物的推动，另一方面则是事物内部各个组件之间的相互作用力导致。

静态和动态分解容易，但是动静结合分析难，即在静态分解完成后，还需要考虑事物本身的动态变化及静态组件之间是如何相互作用推进的。从事物发展演进的动态观来看，事物的各个静态组件本身不是死的，而是活的，正是由于事物动态发展将静态组件之间衔接起来了。

一个软件系统在设计实现的时候会分解为多个组件，但是一个业务流程和功能究竟如何流转？那就存在组件之间要进行交互，即对于拿到的动态业务流程需要去分析，组件之间如何进行交互，从获取到一个需求最终完成闭环并产生一个输出。这个想清楚了后系统的具体实现才能够想清楚。即回到前面一直谈到的一个点，静态分解往往本身不是目的，静态分解后的组件如何交互和集成，完成和实现一个动态的目标，或者说最终完整聚合成最初的事物才是最关键的。

你研究的目标→事物→静态分解→动态分析→动静结合分析→回归目标实现。

任何思维层面的事情、任何分析和决策，都不要忘记了你最初的目标，否则你将会越走越远。特别是在思维中越是分解和分析，越容易不断地深入和陷入细节里。勿忘初心，方得始终，思维中的目标驱动并形成闭环是思维过程中必须时刻意识到的事情。

### 3.1.2 分类

分类是指将事物中各产品按照种类、等级或性质分别归类。

分类有时候也可以叫做归类，用归类可以更加形象化地体现分类的意思。分类是产品分析中的关键方法。当我们面对成千上万的产品时，我们不可能针对成千上万的个体采用不同的决策方法和行动计划，而最可行的方法则是对产品进行分类，通过分类后我们只需要对不同的对象分类采用不同的决策方法和行动计划即可。

面对一个产品群，首先要考虑的是分类，即可以根据产品的哪些关键属性对产品群进行初步的分类整理。拿投资理财的来说，你可能会面对诸多的产品和选择，如果从风险和收益考量你可以初步分类为银行理财产品、信托、债券、基金、P2P理财等几个大类。对于基金类可能又可以分为保守类的债券投资基金和激进的股票基金，而任何一个基金产品本身又是多个最终股票形成的股票池的产品组合。

分类是第一步，在分类后即可以根据事物本身的关键特征和属性，选择关键的两个维度属性进行矩阵分析，如投资理财，风险和收益即可以作为关键的两个矩阵维度。

抽象的目的是研究事物的共性特征，而从具体事物归纳和提炼出抽象的表达，当有了这种抽象的表达后才可能后期对研究的新事物进行演绎和拓展。即抽象的过程是归纳的过程，而解决问题和决策的过程是演绎的过程。

当我们研究公司的一组项目的时候，里面可能包括了研发类项目、销售类项目、公司流程优化项目、运维类项目等，那么这就是一种基本的分类，通过这种分类后即开始研究每类项目应该如何去管理，这类项目本身的关键特征属性究竟是什么。在研究完后我们会发现不论哪种类型的项目，对于计划、任务跟踪、里程碑、汇报机制等项目管理关键属性都是共通的，那么即还需要进一步抽象为公司级的基础项目管理方法论，定义最基本的标准规范体系，最终使公司所有的项目的项目管理能形成一个完整的视图。即各类项目既有共性，也有特殊，共性基础内容必须一致，而特殊内容可以根据项目特征进行定制。

在完成初步的分类后，要研究一个完整的产品群还需要进行进一步的抽象，对于抽象也可以理解为归纳，即找寻不同产品的共同特征和属性，从最终的产品转化为抽象的思维表达。

对于产品的分类，我们需要去找到事物所具备的共有属性，然后根据这些共有属性所表现出的不同内容进行分类的划分。这些属性可能是事物具备的静态属性，如产品是圆形还是方形，也可能是事物表现出来的动态属性，如直线运动还是曲线运动，这些都可以作为对产品进行分类的属性。

以上分析可以看到分解的过程是一个从顶朝下的过程，而分类和抽象的过程恰好是一个从底向上的过程。

分解的目的是由整体到个体，同时通过个体的分析来洞悉事物内在运行机制；

分类的目的则是从个体到整体，通过分类和抽象来实现对抽象类别的统一决策和行动。

因此，不对建筑业信息进行分解的建筑业信息分类必然是个庞大、难以控制的分类。

## 3.2 工作分解结构（Work Breakdown Structure，WBS）

实现数字化表达的第一步是要弄清楚要数字化表达的最终交付成果是什么。试图描绘这个最终交付成果，需要的是工作分解结构。

### 3.2.1 工作分解结构（Work Breakdown Structure, WBS）

WBS 是一种范围管理的工具，最早是由美国国防部提出的。《项目管理知识体系指南》（第 6 版）中将工作分解结构定义为"工作分解结构是项目为实现项目目标、创建所需可交付成果而需要实施的全部工作范围的层级分解。工作分解结构每向下分解一层，代表对项目工作更详细的定义。"在项目管理实践中，项目范围是由 WBS 定义的，完成 WBS，最终要交付的成果也就描绘出来了。WBS 总是处于计划过程的中心，也是制定进度计划、资源需求、成本预算、风险管理计划和采购计划等的重要基础。同时，WBS 也是控制项目变更的重要基础。

创建工作分解结构（图 3-1）是把项目可交付成果和项目工作分解成较小的，更易于管理的组件的过程。

工作（Work）——可以产生有形结果的工作任务；分解（Breakdown）——是一种逐步细分和分类的层级结构；结构（Structure）——按照一定的模式组织各部分。根据这些概念，WBS 有相应的构成因子与其对应：

（1）结构化编码

编码是最显著和最关键的 WBS 构成因子，首先编码用于将 WBS 彻底地结构化。通过编码体系，我们可以很容易识别 WBS 元素的层级关系、分组类别和特性。并且由于近代计算机技术的发展，编码实际上使 WBS 信息与组织结构信息、成本数据、进度数据、合同信息、产品数据、报告信息等紧密地联系起来。

（2）工作包

工作包（Work package）是 WBS 的最底层元素，一般的工作包是最小的"可交付成果"，

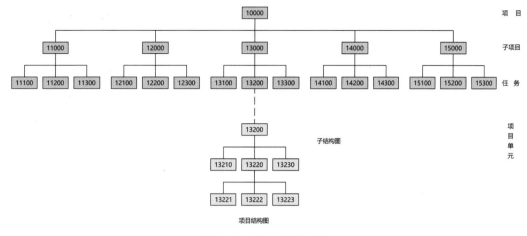

图 3-1　工作分解结构图

这些可交付成果很容易识别出完成它的活动、成本和组织以及资源信息。例如：管道安装工作包可能含有管道支架制作和安装、管道连接与安装、严密性检验等几项活动；包含运输／焊接／管道制作人工费用、管道／金属附件材料费等成本；过程中产生的报告／检验结果等文档；以及被分配的工班组等责任包干信息等。正是上述这些组织／成本／进度／绩效信息使工作包乃至 WBS 成为项目管理的基础。基于上述观点，一个用于项目管理的 WBS 必须被分解到工作包层次才能够使其成为一个有效的管理工具。

（3）WBS 元素

WBS 元素实际上就是 WBS 结构上的一个个"节点"，通俗的理解就是"组织机构图"上的一个个"方框"，这些方框代表了独立的、具有隶属关系／汇总关系的"可交付成果"。经过数十年的总结，大多数组织都倾向于 WBS 结构必须与项目目标有关，必须面向最终产品或可交付成果，因此 WBS 元素更适于描述输出产品的名词组成（effictive WBS，Gregory T. Haugan）。其中的道理很明显，不同组织、文化等为完成同一工作所使用的方法、程序和资源不同，但是他们的结果必须相同，必须满足规定的要求。只有抓住最核心的可交付结果才能最有效地控制和管理项目；另一方面，只有识别出可交付结果才能识别内部／外部组织完成此工作所使用的方法、程序和资源。工作包是最底层的 WBS 元素。

（4）WBS 字典

管理的规范化、标准化一直是众多公司追求的目标，WBS 字典就是这样一种工具。它用于描述和定义 WBS 元素中工作的文档。字典相当于对某一 WBS 元素的规范，即 WBS 元素必须完成的工作以及对工作的详细描述；工作成果的描述和相应规范标准；元素上下级关系以及元素成果输入输出关系等。同时 WBS 字典对于清晰的定义项目范围也有着巨大的规范作用，它使得 WBS 易于理解和被组织以外的参与者（如承包商）接受。在建筑业，工程量清单规范就是典型的工作包级别的 WBS 字典。

创建 WBS 应按照实际工作经验和系统工作的方法、工程的特点、项目管理者的要求进行，其基本原则是：

1）WBS 要遵循 MECE（Mutually Exclusive Collectively Exhaustive）原则，即对一项工作进行分解时，要做到相互独立，完全穷尽；

2）WBS 要遵循 SMART 原则，即一项工作的分解要具备具体（Specific）、可量化（Measurable）、可实现（Attainable）、相关性（Relevant）、有时限（Time-bound）五个条件。每一项工作都必须要有部门和人负责，必须要有主要负责人员，具体到个人，而不是分配给几个人组成的小组；

3）项目的每一个阶段应要能区分不同的责任者和不同的工作内容，应有较高的整体性和独立性；

4）可视化原则，可以分层看到每一项细化的工作；组件可以移动位置以利于编排 WBS；

5）能够符合项目目标管理的要求，能方便地应用工期、质量、成本、合同、信息等手段；

6）WBS 分解层次以 4～6 层为宜，最低层次的工作包的单元成本不宜过大、工期不宜太长。

### 3.2.2　SWBS 与 PSWBS

（1）纲要性工作分解结构（SWBS：Summary WBS）

纲要性工作分解结构是指导性的、战略性的工作分解结构。该分解结构只有上面的三级（图 3-2）：

第一级：建设行业各子行业不同阶段为 SWBS 系统；

第二级：不同阶段下的分系统；

第三级：从属于分系统的子系统，即建筑业建物的基本单元。

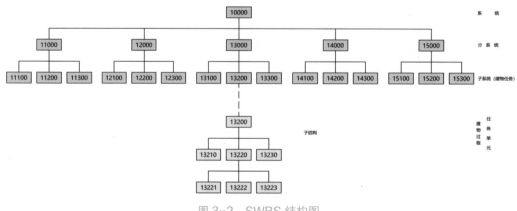

图 3-2　SWBS 结构图

（2）项目纲要性工作分解结构（PSWBS，Project summary work breakdown structure-PSWBS）

项目纲要性工作分解结构是针对某一特定项目，对纲要性工作分解结构进行裁剪得到的工作分解结构。对于具体项目，从图3-2可裁剪得到图3-1的前三级项目纲要性工作分解结构（PSWBS）。

### 3.2.3 体分法

思维是以概念、范畴为工具去反映认识对象的。这些概念和范畴是以某种框架形式存在于人的大脑之中，即思维结构。这些框架能够把不同的范畴、概念组织在一起，从而形成一个相对完整的思想，加以理解和掌握，达到认识的目的。因此，思维结构既是人的一种认知结构，又是人运用范畴、概念去把握客体的能力结构。

逻辑思维（Logical thinking）是指将思维内容联结、组织在一起的方式或形式。它是思维的一种高级形式，是指符合世间事物之间关系（合乎自然规律）的思维方式，我们所说的逻辑思维主要指遵循传统形式逻辑规则的思维方式。常称它为"抽象思维（Abstract thinking）"或"闭上眼睛的思维"。逻辑思维是一种确定的，而不是模棱两可的；前后一贯的，而不是自相矛盾的；有条理、有根据的思维。在逻辑思维中，要用到概念、判断、推理等思维形式和比较、分析、综合、抽象、概括等思维方法，而掌握和运用这些思维形式和方法的程度，也就是逻辑思维的能力。

SWBS是逻辑思维表达。

辩证思维是反映和符合客观事物辩证发展过程及其规律性的思维，是对客观辩证法和认识过程辩证法一定程度的认识和运用。辩证思维的特点是从对象的内在矛盾的运动变化中，从其各个方面的相互联系中进行考察，以便从整体上、本质上完整地认识对象。辩证思维运用逻辑范畴及其体系来把握具体真理。辩证思维既不同于那种将对象看作静止的、孤立的形而上学思维，也不同于那种把思维形式看作是既成的、确定的形式逻辑思维。它是辩证逻辑研究的对象。人类的辩证思维的历史发展经历了一个从自发到自觉的过程。

辩证思维是指以变化发展视角认识事物的思维方式，通常被认为是与逻辑思维相对立的一种思维方式。在逻辑思维中，事物一般是"非此即彼"、"非真即假"，而在辩证思维中，事物可以在同一时间里"亦此亦彼"、"亦真亦假"而无碍思维活动的正常进行。辩证思维指的是一种世界观。世间万物之间是互相联系，互相影响的，而辩证思维正是以世间万物之间的客观联系为基础，进行的对世界进一步的认识和感知，并在思考的过程中感受人与自然的关系，进而得到某种结论的一种思维。辩证思维要求观察问题和分析问题时，以动态发展的眼光来看问题。

图3-1及图3-2中的任务单元或子结构，内含完成任务（建物）的具体活动过程，是属于辩证思维范畴。

体分法是对整体采用"有限线分法"、个体对象结构化编码，将整体以信息完整度较高的不同个体表达的信息分解方法。当我们将体分法应用于 WBS，"有限线分法"将线分法控制在 3 层以内，即采用 SWBS，既保留了信息容量大，层次清楚，逻辑性强，又符合传统应用的习惯；既适合于手工操作，又便于计算机管理，发挥了线分法优点的同时又回避了线分法的所有缺点。

按照体分法，图 3-1 分为两部分，即任务及其之上部分与下部的项目任务单元部分。在建设工程行业，"任务"，即各专业工作和施工工法等基本固定，这部分可采用"逻辑思维"进行分解。其分解是一种确定的，而不是模棱两可的；前后一贯的，而不是自相矛盾的；有条理、有根据的。通过概念、判断、推理等思维形式和比较、分析、综合、抽象、概括等思维方法，使任务及其之上部分合乎自然规律。

图 3-1 的下部分，即"任务单元"部分，"任务单元"是完成任务采用的人机料法环，是个变量，即"任务"的建物过程，应该采用"辩证思维"进行分析。从对象的内在矛盾的运动变化中，从其各个方面的相互联系中进行考察，以便从整体上、本质上完整地认识对象。这些互相联系、互相影响的对象，无论对立统一规律、质量互变规律、否定之否定规律，辩证思维的所有变化规律将在"任务"中得到统一。

## 3.3 模式

### 3.3.1 模式思维

模式没有很好的定义，但是最著名的一个定义是由建筑学家克里斯托弗·亚历山大（Christopher Alexander）给出的"每一个模式描述了一个在我们周围不断重复发生的问题，以及该问题的解决方案的核心。这样，你就能一次又一次地使用该方案而不必做重复劳动"。

根据定义，模式包含两个基本的要素，问题以及解决方案的核心。我们首先来看问题。模式的目的是为了复用问题的解决方案，以减少解决问题的成本，这个前提是问题要能够不断地重复发生。如果问题本身不经常发生，那么问题及其解决方案就不能成为模式。再看模式的另外一个要素，解决方案的核心。为什么要加"核心"两个字呢？这是由于同样一个问题在不同的背景下发生时，其解决方案的核心是相同的，但是具体的实施细节上可能会有差异。然而，如果问题的解决方案的核心才是解决问题的难点，背景差异带来的影响是可以轻易解决的，那么仍然没有违背模式的目的。

模式的关键点是它们来源于实践。必须观察人们的工作过程，发现其中的方案设计逻辑，才能找出"这些解决方案的核心"。

每一个模式相对独立，但又彼此不孤立，它们之间相互影响。模式的价值不在于它能够给我们多少新的内容和形式，而是可以确信无疑地告诉我们哪一些解决方案在实践的过程中是行之有效的，并且模式的思维（我们还会要强调模式来源于实践）可以帮助我们进行知识的复用。

### 3.3.2 模式思维在建筑工程领域的应用

在建设领域我们有经典的符合我国国情的构件划分、计算方法、实施工艺、资源组合方式，这些经典的可复用的解决方案处处都体现了基于实践的模式思维。这些概念都有着相对固定的形态和实施逻辑。这是我们可以进行建筑业信息分解编码的一个基本前提。

IDM 是针对"完全覆盖建筑物全生命期的应用系统中实际工作流程单一任务（软件）所需的交互信息（MVD）"。因此，IDM 是一个系统工程。爱因斯坦说过：我们不能在产生问题的层次上去寻找解决的办法，而应当进入更高层次上去思考。我国建筑业不同子行业有其固定的工作流程，不同工作流程节点需要任务软件。工作流程满足"模式"的定义。

"模式思维"就是结合建筑业工作流程，统筹策划完整的"分布式 BIM 数据库及其 IDM/MVD 体系"，以分布式数据库替代"构件思维"的单一格式（"集中式"）BIM 数据库，以分布式软件系统替代 BIM 建模软件，以端对端多样化标准交换需求替代单一的 IFC 标准交换。形成一种全新的 BIM"模式思维"，这种由若干不同但互相联系功能部件组成的动态 BIM 系统，是一种化整为零、各个击破；化繁为简、做精做全；多方位参与、多角度切入、软件多元化的 BIM 实施方式，蕴含着兵法"分而治之"的奥妙。国家《建筑信息模型应用统一标准》提出的 P-BIM 本质上就是一种模式（PATTERN）思维。

相对于互联网的一日千里，BIM 的"构件思维"发展显得缓慢而保守，现有互联网服务提供的空间能力并非包罗万象，留给构件思维的空间仍然巨大，但如果在思想和技术上继续因循守旧，可能会再次错过建筑业信息化的又一机遇期。过去 20 年，BIM 重点在于"构件思维"的数据建模，这仅仅是一种可视化手段，与空间计算无关。而在大数据时代，利用数据支撑决策才是 BIM 价值所在。BIM 辅助决策的核心是综合与高效，构件思维有先天缺陷成为无解难题。反观基于模式思维的 HIM 矩阵网格这种互联网服务的技术路线，网格是空间计算、是信息承载和计算的基本单元，瓦片地图、空间搜索、实时交通、叫车匹配……无一例外全是网格，因为网格在计算机中记录为统一规则的编码，调用编码进行各种运算，是通用的 IT 方法，可借由各种 IT 优化手段应对 BIM 系统海量数据和海量访问。模式思维包含构件思维。

### 3.3.3 模式应用的两种意义

（1）模式是指从生产经验和生活经验中经过抽象和升华提炼出来的核心知识体系。模式（Pattern）其实就是解决某一类问题的方法论。把解决某类问题的方法总结归纳到理论高度，那就是模式。模式是一种指导，在一个良好的指导下，有助于你完成任务，有助于你做出一个优良的设计方案，达到事半功倍的效果，而且会得到解决问题的最佳办法。

（2）模式还是一种认识论意义上的确定思维方式，是人们在生产生活实践当中经过积累的经验的抽象和升华。简单地说，就是从不断重复出现的事件中发现和抽象出的规律，是解决问题形成经验的高度归纳总结。只要是一再重复出现的事务，就可能存在某种模式。

### 3.3.4 建筑业"系统"模式思维

史蒂文·舒斯特（Steven Schuster）在其《系统思维的艺术》里面对系统有一个定义"所谓系统，就是一个由很多部分组成的整体，各个部分互相之间有联系，作为整体又有一个共同的目的。人的身体、学校、公司、国家，都是系统。"

建筑业信息分解是人们长期实践活动的经典模式系统。传统的基于实践的模式都是离散的存在，并且由于缺乏层次和架构我们很难将其进行逻辑上的组合，很难突破性地利用模式的价值，建筑业信息分解给出了这些模式的实施边界，定义了各个模式间的关系，兼顾了个体和系统间的目标，在这一层面来看，建筑业信息分解体系是一种系统论的应用。

## 3.4 建筑业任务分解体系

建设行业（建筑业）是个庞大系统，建筑业信息化是为建筑业任务服务。只有深入分析建筑业工程技术与管理现状和模式，才能使信息化与工程实践结合，做出有用的信息系统。在工程实践中我们将建筑业分为建筑工程、公路工程、市政工程、铁路工程等子行业，不同子行业间既独立又相互关联。对于每个子行业的项目具有自己独特的 SWBS，SWBS 适用于项目集中的任何项目 PSWBS。建筑业任务分解体系如图 3-3 所示。

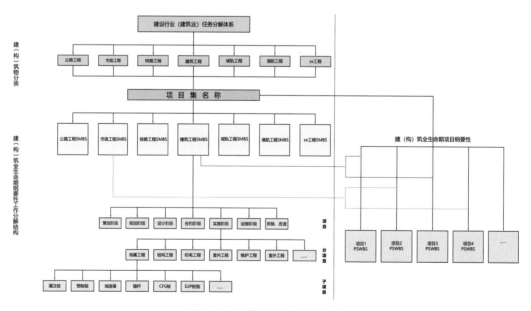

图 3-3 建筑业任务分解体系

图 3-3，首先将建筑业分解为不同子行业，对不同子行业的项目集（多个项目的集合）进行命名，然后再对项目集的共性进行 SWBS 分解，具体项目 PSWBS 则可从 SWBS 中裁剪而成。

将图 3-3SWBS 中的第三层子系统（建物任务）、第二层建物、第一层阶段建物及每项工作对应的组织分解结构（OBS-organizational breakdown structure）名称编码用列矩阵表达为：

$$\left\{\begin{array}{c} \text{SWBS} \\ \text{第三层} \\ \text{建物编码} \\ 11100 \\ 11200 \\ 11300 \\ 12100 \\ 12200 \\ 12300 \\ 13100 \\ 13200 \\ 13300 \\ 14100 \\ 14200 \\ 14300 \\ 15100 \\ 15200 \\ 15300 \\ \cdots\cdots \\ \text{第二层} \\ \text{建物编码} \\ 11000 \\ 12000 \\ 13000 \\ 14000 \\ 15000 \\ \cdots\cdots \\ \text{第一层} \\ \text{建物编码} \\ 10000 \\ \cdots\cdots \\ \text{各层对应} \\ \text{OBS编码} \\ 90000 \\ 91000 \\ 92000 \\ \cdots\cdots \end{array}\right\} \quad (3\text{-}1)$$

## 3.5 建筑信息模型分解结构（模型分解结构，Model Breakdown Structure，MBS）

模型分解结构（MBS）基本定义：以可独立交付成果信息为导向对项目建筑信息模型要素进行分组，它归纳和定义了从项目单一建筑信息模型的主从式数据库结构系统改变为分布式数据库结构系统的每一层级完成项目建筑信息模型工作的更详细定义。

纲要性模型分解结构（SMBS）对应于 SWBS。

数字孪生（Digital Twin）是一个物理产品的数字化表达，以便于我们能够在这个数字化产品上看到实际物理产品可能发生的情况，与此相关的技术包括增强现实和虚拟现实。

数字之索（Digital Thread）在设计与生产的过程中，可以仿真分析模型的参数，传递到产品定义的全三维几何模型，再传递到数字化生产线加工成真实的物理产品，通过在线的数字化检测/测量系统反映到产品定义模型中，进而又反馈到仿真分析模型中。

从辩证思维角度考虑，既然创建 WBS 是把项目交付成果和项目工作分解成较小的、更易于管理的组成部分的过程；那么，创建 MBS 是把整体建筑信息模型（Model）按照 WBS 结果进行分解，将项目信息交付成果和项目信息整体分解成较小的、更易于管理的信息组成部分的过程。通过与工作分解结构的任务层相对应，即 WBS 的数字化表达。

映射：假定 $A$，$B$ 是两个集合，如果按照某种对应法则 $f$，对于集合 $A$ 中的任何一个元素 $x$，在集合 $B$ 中都有唯一的元素 $y$ 和它对应，那么这样的对应叫做集合 $A$ 到集合 $B$ 的映射，记做 $f: A \rightarrow B$。并称 $y$ 是 $x$ 的象，$x$ 是 $y$ 的原象。

美国 BIM 标准首先定义了：BIM 是一个设施物理和功能特性的数字化表达（Model）。式（3-1）中的第三层"建物"与 MBS 分布式数据库完全映射，是最详细的信息定义，满足美国 BIM 标准对 Model 的要求："BIM 是一个设施有关信息的共享知识资源，从而为其全生命期的各种决策构成一个可靠的基础，这个全生命期定义为从早期的概念一直到拆除。"

WBS 归纳和定义了项目的整个工作范围每下降一层代表对项目工作的更详细定义，MBS 则是归纳和定义了项目的整个工作范围信息每下降一层代表对项目工作的更详细信息定义。由于 MBS 是分布式数据库，为保证第三层分布式数据库的协调和一致性，位于任务之上是协同数据库，即式（3-1）中的第二层"分系统建物"与分系统协同系统数据库完全映射；同理，第一层"系统建物"与系统数据库完全映射。

美国 BIM 标准定义："BIM 的一个基本前提是项目全生命期内不同阶段不同利益相关方的协同，包括在 BIM 中插入、获取、更新和修改信息以支持和反映该利益相关方的职责。"

项目的组织分解结构（OBS）是关于项目内部组织的，而不是组织要素与其母体组织或其他机构的关系。它是用与工作分解结构相似的方法构建而成的项目的内部组织图表。组织分解结构描述了负责每个项目活动的具体组织单元，它是一种将工作包与相关部门或单位分层次、有条理地联系起来的项目组织图。OBS 是一种演化而来的方法，负责在项目范围内分解各层次人员的责任。OBS 不等同于内部的组织分解体系，比如一名人员虽然处于比较低的组织体系层次，但他/她可能需要了解全局的，因此就可能需要处于较高的 OBS 层次上。另外，OBS 还包括各项目参与方的组织，甚至可以扩展到各"项目利益利害关系者（Project Stakeholders）"。项目利益利害关系者（业主、政府、设计院、总包、分包……）利用各自项目管理平台获取和创建 BIM 数据，因此而产生 OBS 组织管理对应的数据库，各层对应 OBS 与项目利益相关方的管理数据库映射。

因此，式（3-1）中 SWBS 对应 SMBS 完全映射为：

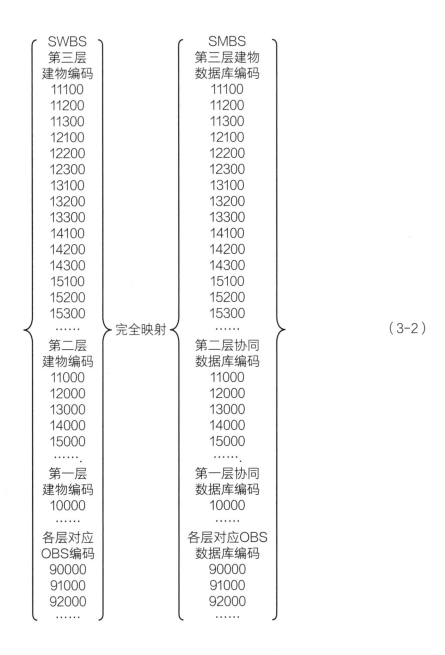

（3-2）

## 3.6　分布式功能建模软件（P-BIM 功能软件）

### 3.6.1　建筑业的非线性特征

因果关系明确的问题都是线性问题——既然这有一个结果，就一定有一个原因，只要解决了原因就能解决问题。举一个简单的例子，使用的手机没有电了，充电就可以了；还比如汽车的轮胎老化了，我们更换一个新的合格的轮胎就可以，线性问题简单明了，因果关系明确。

因果关系复杂或者不明确的问题都是非线性问题，建设工程项目的问题大多是非线性问题，比如一个墩柱的混凝土质量有问题，有可能是出场的混凝土不合格，也有可能是浇筑时候的操作违规（加水或者等待时间过长），也有可能是养护不到位，甚至是出现了极端的反常天气所致。

建筑业系统本质上是不满足叠加原理的不确定非线性系统。

### 3.6.2　传统的非线性系统的建模方法

复杂系统的建模和控制一直是控制界研究的热点问题，传统的全局建模方法普遍存在着待定参数过多，模型结构难以确定等问题，如果所关注的对象是一个高维空间问题，还存在易陷于"维数灾难"的问题。同时多数方法的使用效果与使用者的经验有关。在实际使用中，要想获得高精度的估算，通常需要消耗大量的时间和精力来选择恰当的建模方法，确定合适的模型结构或者模型集合，同时还需要对算法和参数进行反复调整。然而，即使获得了高精度的模型，往往也由于得到的模型过于复杂而难以做进一步的分析和应用。

控制工程领域广泛存在着这样一类复杂非线性系统，由于非线性因素的作用，系统运行模态过于复杂，或者工作环境非常恶劣，造成系统结构和参数随着工况的改变而呈现出很强的不确定性。此时，采用传统的全局建模方法建立单一的高精度数学模型就显得比较困难。因为多数建模方法的使用效果与使用者的经验有很大关系，模型辨识复杂，计算量偏大，耗时偏长，而难以实际应用。有时候，即使得到了系统的模型，也往往由于模型本身过于复杂而难以进行稳定性分析和控制器设计。为了解决该问题，控制界的专家和学者们进行了大量的理论和技术探索，最终发现基于"分而治之"思想的多模型方法是解决问题的一种有效途径。

### 3.6.3　多模型建模理论

多模型方法自 20 世纪 70 年代出现以来，已经走过了 40 余年的发展历程，无论是在理论、技术，还是在应用方面都取得了广泛的成功。该方法以"分解—合成"原理作为解决问题的自然之道，将复杂系统的建模与控制问题按照某种确定性准则进行分解，然后分别建立局部模型和控制器，由于局部模型常常是线性的，因此可利用成熟的线性系统理论进行控制器的设计。在问题分解并获得解答后，即可按照相应的合成准则或调度机制协调各局部模型或控制器，在保证稳定性的同时，实现对原复杂非线性系统的控制。这种建模与控制方法能够将复杂的非线性问题转化为可用成熟且相对简单的理论来解决的问题，这对简化控制器设计、提高系统在复杂环境下的控制性能具有重大意义。当前，多模型方法主要有多模型建模方法、多模型控制方法和交互式多模型滤波方法等，但无论哪类方法，需要解决的关键问题无外乎两点：一是在特定的分解准则下选择恰当的模型集；二是选择恰当的合成准则确保模型调度的稳定性。

多模型建模理论最重要的方法论就是"分解—合成"，通过对"分解—合成"建模方法的应用，将整个非线性系统进行分解，用若干个简单的局部模型实现对原有系统的逼近。这种建模框架具有两大优点：

（1）能够利用成熟的线性系统建模策略分析每一个局部模型的特征；

（2）对复杂的非线性系统，必须选择复杂的模型结构来逼近整个系统，只要采用合适的学习策略和补偿机制，既可以达到合理的逼近精度，也可以降低计算的复杂性。

为此，人们提出利用多个局部模型来逼近系统的动态特征，并基于每一个局部模型设计自适应控制器的思想。多模型方法具有智能控制的特点，能够把经典的建模、控制方法与先进的控制思想结合起来，基本原理简单，算法简便，且易于实现。

分段防射/线性多模型这种方法将状态的空间分割成有限个面角区域，在每一个面角区域用一个分段防射/线性函数来描述系统的动态特征。基于多模型"分解—合成"的思想，分段防射/线性多模型将系统的空间分割成有限个子区间，在每一个子区间上用一个分段防射/线性局部模型来表述，然后通过切换的方式来描述系统的动态。这类模型除了可以任何精度逼近任意一个足够平滑的非线性函数外，还可以用来逼近具有非连续特性的非线性系统。另外，它还与混杂系统密切相关，并被证明完全等价于混合逻辑动态系统、线性互补系统、拓展线性互补系统以及最大最小正坐标量系统等结构类的混杂系统。因此，分段防射/线性多模型可以用于混合系统的辨识。

求解分段防射/线性系统辨识问题最直接的方法是，根据模型结构以及参数选择代价函数，然后直接采用数值方法或优化算法进行求解。

随着被控制系统对象的日益复杂，越来越多的现代系统呈现出强非线性、强耦合、工况范围广等特点，采用单一模型对全局系统进行描述常常会因为建模结果过于复杂，难以进行控制器设计而无法满足实际的需求。另外，即使能够获得满意的全局模型。其辨识过程也可能因为算法本身过于复杂，所需数据规则太多，造成计算量偏大、耗时偏长，而难以在现实中应用。因此，从本控制对象的样本数据出发，充分挖掘其中隐藏的有用信息，寻找其他有效的建模方法成为必要。

"合而治之"思想的不成功促使人们转向"分而治之"，用被控制对象的多个局部模型代替单一全局模型分别进行建模，最后将局部多模型通过某种手段综合为一个整体来逼近全局模型，这就是非线性系统建模的"分解—合成"原理。基于"分解—合成"原理的多模型建模方法可简要描述为 3 个步骤：

首先，按照某种分解准则，将整个系统工作区域划分为若干个区间，定义全区间以及用来表征区间的特征变量；

其次，选取各个区间上的局部模型结构，并辨识其参数；

最后，根据某种调度机制或者性能指标将各局部模型进行组合得出原有问题的解。

SMBS 就是基于多模型建模思想，软件是辅助完成任务的工具。对于 WBS 而言，每项任务必然存在一个应用软件（在 WBS 系统中对应软件功能）；而对于 SMBS 分布式数据库而言，每个数据库必然有一个建模软件，因此，分布式功能建模软件（P-BIM 软件功能）编码与 WBS 编码和 MBS 编码完全映射，见式（3-3）。

$$
\left.\begin{array}{l}
\left\{\begin{array}{l}
\text{SWBS} \\
\text{第三层} \\
\text{建物编码} \\
11100 \\
11200 \\
11300 \\
12100 \\
12200 \\
12300 \\
13100 \\
13200 \\
13300 \\
14100 \\
14200 \\
14300 \\
15100 \\
15200 \\
15300 \\
\cdots\cdots \\
\text{第二层} \\
\text{建物编码} \\
11000 \\
12000 \\
13000 \\
14000 \\
15000 \\
\cdots\cdots \\
\text{第一层} \\
\text{建物编码} \\
10000 \\
\cdots\cdots \\
\text{各层对应} \\
\text{OBS编码} \\
90000 \\
91000 \\
92000 \\
\cdots\cdots
\end{array}\right\} \text{完全映射}
\left\{\begin{array}{l}
\text{SMBS} \\
\text{第三层建物} \\
\text{数据库编码} \\
11100 \\
11200 \\
11300 \\
12100 \\
12200 \\
12300 \\
13100 \\
13200 \\
13300 \\
14100 \\
14200 \\
14300 \\
15100 \\
15200 \\
15300 \\
\cdots\cdots \\
\text{第二层协同} \\
\text{数据库编码} \\
11000 \\
12000 \\
13000 \\
14000 \\
15000 \\
\cdots\cdots \\
\text{第一层协同} \\
\text{数据库编码} \\
10000 \\
\cdots\cdots \\
\text{各层对应OBS} \\
\text{数据库编码} \\
90000 \\
91000 \\
92000 \\
\cdots\cdots
\end{array}\right\} \text{完全映射}
\left\{\begin{array}{l}
\text{P-BIM} \\
\text{第三层建物} \\
\text{建模软件编码} \\
11100 \\
11200 \\
11300 \\
12100 \\
12200 \\
12300 \\
13100 \\
13200 \\
13300 \\
14100 \\
14200 \\
14300 \\
15100 \\
15200 \\
15300 \\
\cdots\cdots \\
\text{第二层协同} \\
\text{建模软件编码} \\
11000 \\
12000 \\
13000 \\
14000 \\
15000 \\
\cdots\cdots \\
\text{第一层协同} \\
\text{建模软件编码} \\
10000 \\
\cdots\cdots \\
\text{各层对应OBS} \\
\text{建模软件编码} \\
90000 \\
91000 \\
92000 \\
\cdots\cdots
\end{array}\right\}
\end{array}\right. \quad (3\text{-}3)
$$

## 3.7 P-BIM 功能软件信息交换标准

美国 BIM 标准对 BIM 的基本要求是：BIM 是基于协同性能公开标准的共享数字表达。每一个 P-BIM 功能软件都必须配以一本《××P-BIM 软件功能与信息交换标准》，如《建筑设计 P-BIM 软件功能与信息交换标准》对应于 openBIM 所有有关建筑设计软件交付给其他软件的 MVD，如图 3-4 所示。

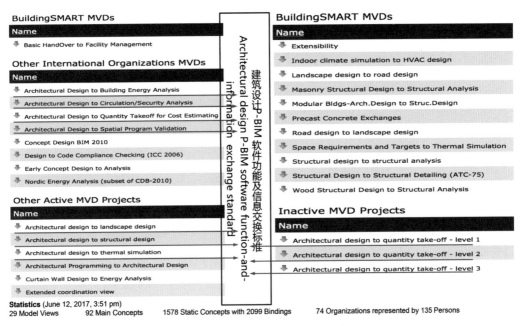

图 3-4　P-BIM 交换标准与 MVD 对照图

因此，功能建模软件交换标准编码也与 P-BIM 软件功能对应，形成 SWBS/SMBS/P-BIM 软件功能/P-BIM 软件信息交换标准编码的完全映射关系，如式（3-4）所示。

## 3.8 建筑业信息分解编码体系（A&bCode）

式（3-4）给出了基于 SWBS 编码，对应了 SMBS、P-BIM 软件功能、P-BIM 软件信息交换标准编码，建筑业信息化是为建筑业基本工作单元服务，如图 1-4 所示，信息交换目标是 Exchange supporting an action or business，实际上 P-BIM 实施体系，每项 Action，如灌注桩施工（图 3-5），一般都包含了 business，故以 Action and business（A&b）表达更为合适。

A&b 单元是工程全生命期不同阶段的最小单元，是 WBS 节点或 MBS 中一单元，可

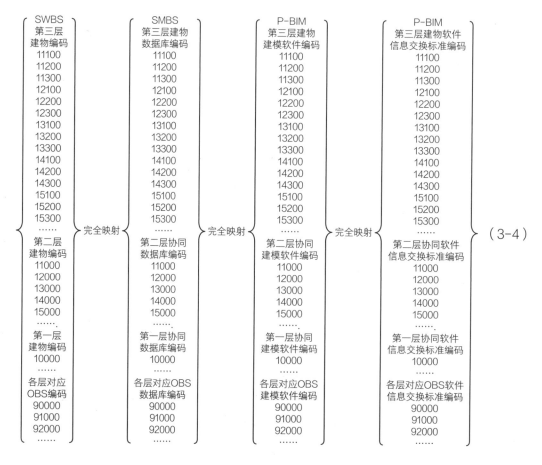

（3-4）

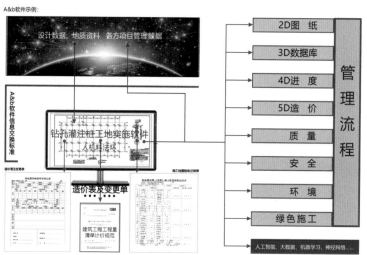

图 3-5 灌注桩施工 A&b 编码软件

执行；第三级 A&b 单元可纵向再分解但包含于 A&b 软件中，因此，A&bCode 分解与各国工程管理流程、工程技术标准及工作习惯无关。A&b 使流程、组织结构（责任分工和职能分工)、资源分解结构及造价体系浑然一体，具备了工程建设项目的所有技术与管理要素。

因此，我们可以 A&bCode 代表式（3-4）具体数值，称之为建筑业信息分解编码体系（A&bCode）。

每个 A&bCode 由四位阿拉伯数字和一个英文小写字母组成，表达为 XYMMa，如 2125a。首位数 X 代表项目全生命期的不同阶段，第 2 位数 Y 代表不同阶段 WBS 分解的"子项目"，第 3、4 位数 MM 代表子项目中的具体工作任务，最后一位字母表示建设行业某一子行业（如建筑工程位 a、公路工程位 b……）。

按照体分法的建筑业信息分解编码如表 3-1 所示。

建筑业信息分解编码（A&bCode） 表 3-1

| A 建筑工程 |
| --- |
| 　　1000a 项目策划 |
| 　　2000a 项目规划 |
| 　　2001a 规划和报建 |
| 　　2002a 规划审批 |
| 　　3000a 项目设计 |
| 　　3001a 总图设计 |
| 　　3002a 建筑设计 |
| 　　3100a 地基设计 |
| 　　3101a 岩土工程勘察 |
| 　　3102a 基坑工程设计 |
| 　　3103a 地基处理设计 |
| 　　…… |
| B 公路工程 |
| C 铁路工程 |
| D 港口与航道工程 |
| E 水利水电工程 |
| F 电力工程 |
| G 矿山工程 |
| H 冶金工程 |
| I 石油化工工程 |
| J 市政公用工程 |
| K 通信工程 |
| L 机场工程 |
| M 核工程 |

对照图 3-1 的建筑业任务分解体系得出建筑业信息分解编码体系架构如图 3-6 所示。

图 3-6 的建筑业信息编码体系的架构逻辑如下：

首先建立一个《建设行业（建筑业）信息分解编码统一标准》，这个标准的意义是对

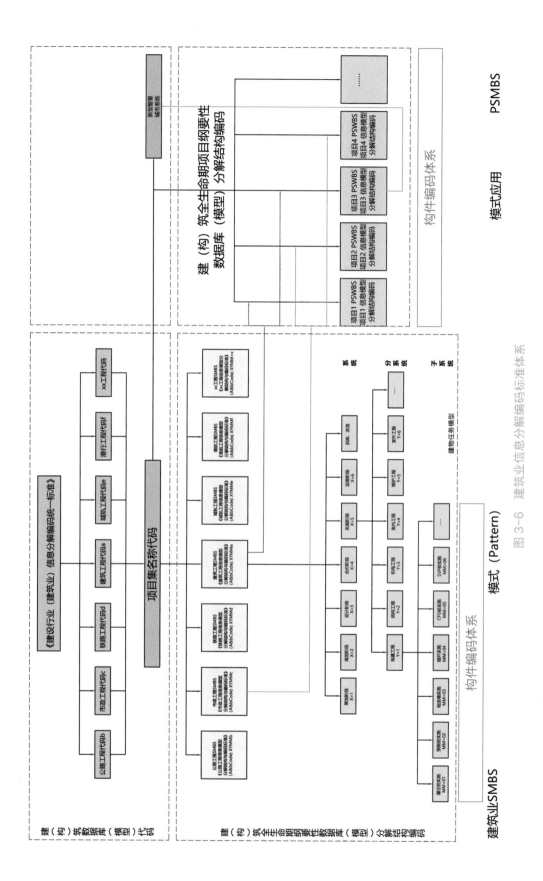

图 3-6 建筑业信息分解编码标准体系

建筑业的体系思维,将建筑业整体分解形成不同子行业项目集的编码。这是数字中国或智慧中国的基本编码体系。

其次,在项目集的编码之下,建立一种可复用的不同子行业 SWBS 编码系统模式,如《公路工程信息模型分解结构与编码标准》《建筑工程信息模型分解结构与编码标准》《市政工程信息模型分解结构与编码标准》《铁路工程信息模型分解结构与编码标准》《机场工程信息模型分解结构与编码标准》等。

这两个层次的分解编码构成了我国建筑业信息分解编码体系架构。即由《建筑业信息分解编码统一标准》确定建筑业所有项目集编码;再由《××子行业信息模型分解结构与编码标准》,如《建筑工程信息模型分解结构与编码标准》,确定各子行业项目集的模型分解模式以应用于任何具体项目。

由上述可知,建筑业信息分解编码体系的关键在于子行业的信息模型分解结构及其编码标准,切合实际的子行业信息模型分解结构应该是以 SWBS 为起点、形成 SMBS、并配以建模软件及其信息交换标准。

A&bCode 是基于逻辑思维的结果,使 BIM 实施模式化。

## 3.9　A&bCode 核心思想及其意义

A&bCode 的核心思想是以终为始、以人为本,不管采用什么样逻辑的方式(线分类法、面分类法、混合分类法、体系分类法、聚类法)来对建筑业信息进行分类,分类到最后的最小值也就是底层都会指向了同一件事物——人(使用者)(如图 3-7 所示),人是推动建设行业也是制约建设行业发展的最大因素。因此信息分类应该是为人服务的,应该符合使用者的逻辑思维。

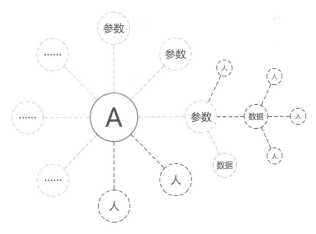

图 3-7　人是最底层的因数

A&bCode 不仅适合我国工程建设信息化，还可以国际化，其原因在于它是业务流的表现和定义，业务流可以根据各个国家不同作业形式补充定义，A&bCode 是业务流最小颗粒度可视化划分，和 OmniClass 表 23 融合且包容，A&bCode 负责过程，OmniClass 表 23 负责结果。A&bCode 是 BIM 不依赖于 "BIM 建模软件"创新发展的建筑业信息化驱动器。

基于建筑信息模型分解结构（MBS）的 A&bCode 分类编码优点在于：

（1）A&bCode 抽象描述了建筑业体系工程（系统的系统）的各要素；

（2）A&bCode 具象描述了符合工程全生命期管理要求的可交付实体工作内容；

（3）A&bCode 与人有关，每个项目参与者工作都能与 A&bCode 单元对应；

（4）A&b 软件（P-BIM 功能软件）内置不同国家可交付实体技术、管理流程、造价和最终构件产品，因此，A&bCode 编码与国家无关；

（5）A&bCode 只对 WBS 工作节点单元编码，各国确定的第三级工作节点基本相同，如有不同可以增列；

（6）A&bCode 只对工作节点软件功能编码，不对程序内容编码，不干涉节点从业者习惯；

（7）A&bCode 是 BIM 系统顶层设计，不影响软件开发者的功能软件创新；

（8）A&bCode 目标是数据交换，有利于软件开发者的功能软件增值；

（9）A&bCode 为项目参与各方创造了无附加条件的协同工作生态环境。

## 3.10　A&bCode 编码标准制定示范（公路工程）

公路工程全生命期按照项目实施的先后顺序，分为策划、规划、勘察设计、合约（招标投标）、实施（竣工）、运维六个阶段，各阶段内工作及管理任务相对独立，并存在项目周期链条的上下游关系。各阶段的 WBS 划分本着建造工程实体（建物）的原则，"建物"在不同的阶段表现形式有所区别，可以是施工图纸、施工合同、隧道分项工程等，但都是为了完成项目的实体而做的任务工作。

### 3.10.1　公路工程 SWBS 分解方法

公路工程工作分解结构应以建造工程实体为原则，根据不同阶段的工作任务和实际工作内容，进行 WBS 分解，如：设计阶段首先分解为：路线设计、路基路面设计、桥梁设计、隧道设计、机电设计、房建设计、环保设计、交安设计等；合约阶段与实施（竣工）阶段 WBS 分解完全一致，首先分解为：临时工程、路基工程、路面工程、桥梁工程、隧道工程、安全设施及预埋管线工程、绿化及环境保护设施、建筑工程、机电工程。然后按照现有施工工艺，结合工程量清单、定额和质量检验评定标准等规范和标准进行分解，具体路基工程又分为：土方路基、石方路基、挡土墙、排水沟等。在公路工程建设的全生命期，同一状态的工作分解结构原则应具有唯一性，且各阶段 WBS 均不宜超过 3 级。

### 3.10.2　公路工程 A&b Code 编码

公路工程全生命期编码是基于各阶段 WBS 分解结果进行编码，采用四位数字 + 一位小写字母（b 代表公路工程）。其中四位数字中左起第一位，代表建设工程的实施阶段划分，编码意义如下："1"代表策划阶段；"2"代表规划阶段；"3"代表设计阶段；"4"代表合约阶段；"5"代表实施竣工阶段；"6"代表运维阶段；"0"专为各相关方管理提供识别码。其中四位数字中左起第二位，代表建设工程的专业或工作面划分，代码意义应视各工程领域的实际情况而定，公路工程第二位代码意义如下："1"代表临时工程；"2"代表路基工程；"3"代表路面工程；"4"代表桥涵工程；"5"代表隧道工程；"6"代表安全设施及预埋管线；"7"代表绿化及环境保护设施；"8"代表建筑工程，参照建设工程执行；"9"代表机电工程。其中四位数字中左起第三、四位，代表建设工程在其专业或工作面下的具体工作内容，代码意义应视各工程领域的实际工作内容而定，定义每一项具体工作的任务有唯一的编码与之对应。每个公路工程 A&b 中涉及的信息分类编码可参照交通运输部《公路工程信息模型应用统一标准》，待有其他方面的分类编码方式时可替换。

### 3.10.3　公路工程 WBS 编制依据

（1）按照现有施工工艺，结合工程量清单、定额和质量检验评定标准等规范和标准编制。

（2）造价信息按照各阶段 WBS 对应造价信息编写。规划阶段对应《公路工程估算指标》，设计阶段对应《公路工程概算定额》《公路工程预算定额》，招投标阶段对应《公路工程预算定额》《公路工程标准施工招标文件（工程量清单）》，施工阶段对应《公路工程预算定额》《公路工程施工定额》，运维阶段对应《公路工程养护定额》。

（3）质量信息为《公路工程标准施工招标文件》第七章技术规范条目号、《公路工程质量检验评定标准》的相关要求和公路工程质量检验表格等。

（4）责任人信息按照行业实际参与方岗位名称及相关规定编写。

### 3.10.4　公路工程 A&bCode 编码标准编制示例

如图 3-8、图 3-9 所示。

### 3.10.5　公路工程管理系统

（1）P-BIM 功能软件

P-BIM 功能软件是为满足项目全生命期各阶段 WBS 具体工作需要的软件功能，主要

公路工程A&bCode标准招投标（合约）阶段WBS编制工作样表

| 序号 | 路基工程WBS（三级） | | | 全生命期A&b编码 | 公路工程信息模型统一编码 | 施工工艺 | 造价信息 | | | 质量信息 | | 安全信息 | 责任人信息 |
|---|---|---|---|---|---|---|---|---|---|---|---|---|---|
| | | | | | | | 预算定额编号 | 工程量清单编号 | 工程量清单计量规则 | 招标文件（技术规范）对应条款 | | | |
| 1 | 2 | 3 | 4 | 5 | 6 | 7 | 8 | 9 | 10 | 11 | | 13 | 14 |
| 1 | | 路基土石方 | | 4201B | | | | | | | | | |
| 2 | | | | 4202B | | | | | | | | | |
| 3 | 路基工程 | 排水工程 | 排水沟 | 4203B | | 人工挖排水沟 | 1-2-1 | 203-1-a | 第207节表207 | 207.04.3 | | | 招标人、投标人、招标代理 |
| 4 | | | | | | 石砌排水沟 | 1-2-3 | 207-2-a、207-2-b、207-2-f | 第207节表207 | 207.04.3 | | | 招标人、投标人、招标代理 |
| 5 | | | | | | 混凝土排水沟 | | | | | | | |
| 6 | 全生命期协同软件及项目管理平台 | 协同软件 | 招投标（合约）软件 | | | | | | | | | | |

图 3-8　公路工程 A&bCode 标准招投标（合约）阶段 WBS 编制工作样表

公路工程A&bCode标准实施（竣工）阶段WBS编制工作样表

| 序号 | 路基工程WBS（三级） | | | 全生命期A&b编码 | 公路工程信息模型统一编码（附录H） | 施工工艺 | 造价信息 | | | 质量信息 | | | 安全信息 | 责任人信息 |
|---|---|---|---|---|---|---|---|---|---|---|---|---|---|---|
| | | | | | | | 工程量清单编号 | 预算定额编号 | 施工定额编号 | 招标文件（技术规范）对应条款 | 质量检验评定标准对应条目 | 质量检验验收评定对应表格 | | |
| 1 | 2 | 3 | 4 | 5 | 6 | 7 | 8 | 9 | 10 | 11 | 12 | 13 | 14 | 15 |
| 1 | | 路基土石方 | | 5201b | | | | | | | | | | |
| 2 | | | | 5202b | | | | | | | | | | |
| 3 | 路基工程 | 排水工程 | 排水沟 | 5203b | | 人工挖排水沟 | 203-1-a | 1-2-1 | | 207.04.3 | 5.5.1 5.5.2 5.5.3 | | | 项目经理、总工程师、造价工程师、路基工程师、施工员、质检员、试验员、安全员 |
| 4 | | | | | | 石砌排水沟 | 207-2-a、207-2-b、207-2-f | 1-2-3 | | 207.04.3 | 5.6.1 5.6.2 5.6.3 | | | 项目经理、总工程师、造价工程师、路基工程师、施工员、质检员、试验员、安全员 |
| 5 | | | | | | 混凝土排水沟 | 207-2-c、207-2-d、207-2-e | 1-2-4 | | | | | | 项目经理、总工程师、造价工程师、路基工程师、施工员、质检员、试验员、安全员 |
| 6 | 全生命期协同软件及项目管理平台 | 单位工程协同软件 | 施工软件 | | | | | | | | | | | |

图 3-9　公路工程 A&bCode 标准实施（竣工）阶段 WBS 编制工作样表

功能是根据不同阶段业务需求包括但不限于数据库管理、信息管理、进度控制、成本控制、质量控制、安全管理、环保以及文明施工、劳务分包合同管理、人员管理、材料管理、设备管理、变更管理、施工模拟等。

A&bCode 编码也是软件功能的编码，软件的功能和数据互用应能满足各相关方协同工作，信息共享的要求，应具有数据导入、专业检查、成果交付和数据交付的功能，并满足数据开放性的要求。

（2）协同软件

协同软件的目的是实现各阶段、各子项目不同功能软件之间的数据交换，以及各阶段功能软件与业主方项目管理平台软件、各专业分包方项目管理平台软件，项目各方管理平台软件之间的数据交换，可以生成各阶段不同子项目的业务和管理数据库。

（3）公路工程管理系统

公路工程管理系统包括项目全生命期各方管理平台，如：交通运输主管部门、安监管理部门、造价管理部门、土地管理部门、水利管理部门、环保管理部门、消防管理部门、审计部门、业主单位、设计单位、施工单位、监理单位、工程咨询单位等所有相关单位的管理系统和平台，需要功能软件和协同软件共同支撑。

# 第 4 章　基于 A&bCode 的 HIM 实现互操作性

## 4.1　数字索网

### 4.1.1　开放式数控系统

IEEE（美国电气电子工程师协会）关于开放式系统的定义是：能够在多种平台上运行，可以和其他系统互操作，并能给用户提供一种统一风格的交互方式。通俗地说，开放式数控系统允许用户根据自己的需要进行选配和集成，更改或扩展系统的功能迅速适应不同的应用需求，而且组成系统的各功能模块可以来源于不同的部件供应商并相互兼容。

（1）美国的 NGC 计划

美国是开放式数控系统的发起人，于 1987 年提出了 NGC（Next Generation Workstation/Machine Controller）计划。NGC 计划的目的是为基于开放式系统体系结构的下一代机械制造控制器提供一个标准，这种体系结构允许不同的设计人员开发可相互交换和相互操作的控制器部件。例如多个装置间的协调，装置的全独立编程，基于模型的处理，自适应路径策略和大范围的工作站及实时特性等。NGC 的体系结构是在虚拟机械的基础上建立起来的，通过虚拟机械把系统和模块链接到计算机平台上。

（2）欧洲的 OSACA 计划

OSACA 计划是 1990 年有欧共体国家的 22 家控制器开发商、机床生产商、控制系统集成商和科研机构联合开发的。OSACA 计划提出的"分层的系统平台 + 结构化的功能单元"的体系结构，该体系结构保证了各种应用系统与操作平台的无关性及相互间的互操作业，保证了开放性。

（3）日本的 OSEC 计划

日本的 OSEC 计划，由东芝机器公司、丰田机器厂和 Mazak 公司三家机床制造商和日本 IBM、三菱电子及 SML 信息系统公司共同组建。其目的是建立一个国际性的工厂自动化控制设备标准。在硬件方面，OSEC 计划采用 PC+ 控制卡的结构，有利于层次化、模块化、灵活配置的实现。OSEC 将功能单元分组并结构化在一些功能层中，其开放体系结构包括了 3 个功能层共 7 个处理阶层。

与国际先进水平相比，国内的开放式数控系统的研究还处于初级阶段。

### 4.1.2 Digital Thread（数字之索）

Digital Twin（数字双胞胎）描述的是通过 Digital Thread（数字之索）连接的各具体环节的模型。Digital Thread 通过先进的建模与仿真工具建立一种技术流程，提供访问、综合并分析系统寿命周期各阶段数据的能力，使各部门能够基于高逼真度的系统模型，充分利用各类技术数据、信息和工程知识的无缝交互与集成分析，完成对项目成本、进度、性能和风险的实时分析与动态评估。数字线索具备"全部元素建模定义、全部数据采集分析、全部决策仿真评估"的特点，能够量化并减少系统寿命周期中的各种不确定性，实现需求的自动跟踪、设计的快速迭代、生产的稳定控制和维护的实时管理。在全生命期内，各环节的模型都能及时进行关键数据的双向同步和沟通，基于这些形成状态统一、数据一致的模型，从而可动态、实时评估系统当前和未来的功能及性能。

简单地说，Digital Thread 贯穿了产品全生命期，尤其是从产品设计、生产、运维的无缝集成；而 Digital Twin 更像是智能产品的概念，它强调的是从产品运维到产品设计的回馈。

### 4.1.3 数字索网

OmniClass 虽然是一个系统科学的分类编码体系，但其本身还不够完善，在具体应用中还存在很多问题。OmniClass 试图建造一个无所不知的上帝，想要把所有的属性分类放在这里，希望用户用什么自己去挑。结果导致编码体系纷繁复杂，且始终无法涵盖所有的分类，最终无法落地应用。

当此路修建不通时，我们就要去找原因。可能是建造者的问题，也可能是建造方向的问题。换了这么多编码组织，动用了这么多人力，依然没有解决好问题。如采用逆向思维，探究我们最终想要的是什么，一个"信息完整、按需交付、协同创建、基于共享标准"结构化集成数据库。构建建筑业的工程实体和过程业务的结构化数据库，是工程和业务的信息集成，是工程和业务的数字化表达。

横看成岭侧成峰，远近高低各不同。不识庐山真面目，只缘身在此山中。换个思路，换个方向，就是另一片天地。A&bCode，建筑业信息分解编码，是一套对建设行业（建筑业）信息分门别类以直接应用于工程建设项目信息化工作的信息分解系统。A&bCode 是对 OmniClass 逻辑结构的总结。

对于建设工程而言，项目是由多个不同产品组成，不仅要将经 WBS 分解的数字孪生（Digital Twin）从产品设计、生产、运维的无缝集成，还要实现不同产品之间协同建造，因此，制造业的数字之索（Digital Thread）在建筑业必须是数字索网（Digital thread net）。

因此，综合开放式数控系统和 Digital Thread（数字之索）的概念，以此构建基于

A&bCode 数字索网 HIM，类似于开放式数控系统概念的建筑业数据交换协同平台，来实现 BIM 的互操作性。

## 4.2 基于 A&bCode 的 HIM 数字索网

### 4.2.1 $n$ 次超静定结构的力法方程

对于 $n$ 次超静定结构，用力法计算时，可去掉 $n$ 个多余约束得到力法基本结构，在去掉的 $n$ 个多余约束处代以多个约束力结合原荷载形成力法基本体系。当原结构体系去掉多余约束处的位移为零时，相应的也就有 $n$ 个已知的位移条件建立 $n$ 个关于求解多余约束力的方程：

$$\begin{aligned}
\delta_{11}X_1 + \delta_{12}X_2 + \cdots + \delta_{1n}X_n + \Delta_{1P} &= 0 \\
\delta_{21}X_1 + \delta_{22}X_2 + \cdots + \delta_{2n}X_n + \Delta_{2P} &= 0 \\
&\cdots\cdots \\
\delta_{n1}X_1 + \delta_{n2}X_2 + \cdots + \delta_{nn}X_n + \Delta_{nP} &= 0
\end{aligned} \tag{4-1}$$

式（4-1）可以写成矩阵表达式：

$$\begin{bmatrix} \delta_{11} & \delta_{12} & \cdots & \delta_{1n} \\ \delta_{21} & \delta_{22} & \cdots & \delta_{2n} \\ \cdots & & \ddots & \cdots \\ \delta_{n1} & \delta_{n2} & \cdots & \delta_{nn} \end{bmatrix} \begin{Bmatrix} X_1 \\ X_2 \\ \cdots \\ X_n \end{Bmatrix} = \begin{Bmatrix} -\Delta_{1P} \\ -\Delta_{2P} \\ \cdots \\ -\Delta_{nP} \end{Bmatrix} \tag{4-2}$$

式（4-2）左边成为柔度系数矩阵，$\delta_{ij}$ 称为位移影响系数或柔度系数（图 4-1）。

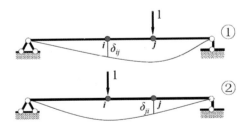

图 4-1  $\delta_{ij}$ 等于 $F_{Pj}=1$ 所引起的与 $F_{Pi}$ 相应的位移

其中 $\delta_{ij}$，下标 $i$ 表示产生位移的点，是位置；

下标 $j$ 表示单位力的作用点，是原因。

### 4.2.2 基于 A&bCode 的数字索网 HIM

按照式（4-2）相同的概念，我们可以利用 A&bCode 建立 BIM 的数字索网 HIM。

如果我们将式（4-2）的柔度系数 $\delta_{ij}$ 理解为 $F_{Pj}=1$（广义"单位"软件）所交付给另一软件 $F_{Pi}$ 相应数据库，那么一个工作节点集成数据库 $HIM_i$ 的矩阵表达式为：

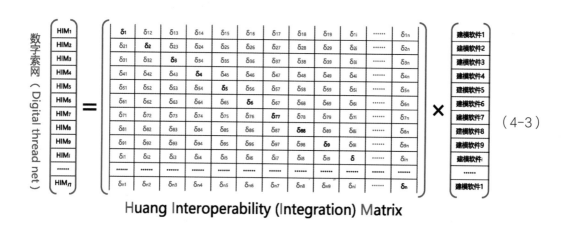

取 $A\&b_{ij}=\delta_{ij} \times$ 建模软件 $_j$，则式（4-3）可表达为：

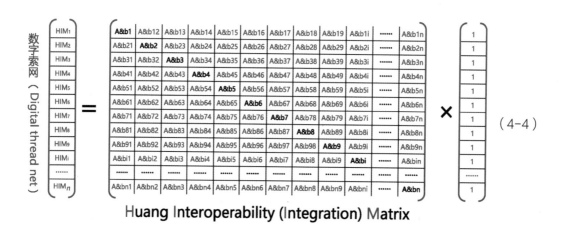

式（4-4）的物理意义如图 4-2 所示。

式（4-4）给出了软件互操作的交换数据库文件夹，建模软件 $i$ 的交换文件夹内容与格式还要通过标准确定。因此，对于不同的建模软件（软件功能）需配套相应的信息交换标准，式（4-4）应进一步表达为：

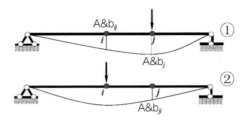

图 4-2 软件数据互操作的"位移影响"表达图

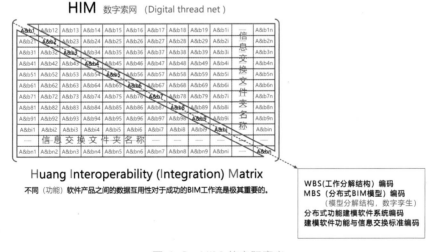

由于 A&bCode 集 WBScode、MBScode、软件功能 code、软件功能信息交换标准 code 为一体，HIM 矩阵实际意义如图 4-3 所示。

图 4-3 HIM 的实际意义

HIM 的简化表达如图 4-4 所示。

$$\begin{pmatrix} A\&b1 & A\&b12 & A\&b13 & A\&b14 & A\&b15 & A\&b16 & A\&b17 & A\&b18 & A\&b19 & A\&b1i & \cdots & A\&b1n \\ A\&b21 & A\&b2 & A\&b23 & A\&b24 & A\&b25 & A\&b26 & A\&b27 & A\&b28 & A\&b29 & A\&b2i & \cdots & A\&b2n \\ A\&b31 & A\&b32 & A\&b3 & A\&b34 & A\&b35 & A\&b36 & A\&b37 & A\&b38 & A\&b39 & A\&b3i & \cdots & A\&b3n \\ A\&b41 & A\&b42 & A\&b43 & A\&b4 & A\&b45 & A\&b46 & A\&b47 & A\&b48 & A\&b49 & A\&b4i & \cdots & A\&b4n \\ A\&b51 & A\&b52 & A\&b53 & A\&b54 & A\&b5 & A\&b56 & A\&b57 & A\&b58 & A\&b59 & A\&b5i & \cdots & A\&b5n \\ A\&b61 & A\&b62 & A\&b63 & A\&b64 & A\&b65 & A\&b6 & A\&b67 & A\&b68 & A\&b69 & A\&b6i & \cdots & A\&b6n \\ A\&b71 & A\&b72 & A\&b73 & A\&b74 & A\&b75 & A\&b76 & A\&b7 & A\&b78 & A\&b79 & A\&b7i & \cdots & A\&b7n \\ A\&b81 & A\&b82 & A\&b83 & A\&b84 & A\&b85 & A\&b86 & A\&b87 & A\&b8 & A\&b89 & A\&b8i & \cdots & A\&b8n \\ A\&b91 & A\&b92 & A\&b93 & A\&b94 & A\&b95 & A\&b96 & A\&b97 & A\&b98 & A\&b9 & A\&b9i & \cdots & A\&b9n \\ A\&bi1 & A\&bi2 & A\&bi3 & A\&bi4 & A\&bi5 & A\&bi6 & A\&bi7 & A\&bi8 & A\&bi9 & A\&bi & \cdots & A\&bin \\ \cdots & \cdots & \cdots & \cdots & \cdots & \cdots & \cdots & \cdots & \cdots & \cdots & \cdots & \cdots \\ A\&bn1 & A\&bn2 & A\&bn3 & A\&bn4 & A\&bn5 & A\&bn6 & A\&bn7 & A\&bn8 & A\&bn9 & A\&bni & \cdots & A\&bn \end{pmatrix}$$

图 4-4　HIM 简化表达

回顾从 IFC 第一版发布至今 20 年的 BIM 发展历程，BIM 的发展是该到了从"构件思维"转向"模式思维"、从"阳春白雪 BIM"转向"大众 BIM"自我革新的时候了。业界常说："软件定义世界，数据驱动未来"，BIM 也是如此。BIM 是由众多独立软件子系统组成的系统，其中，数据驱动软件，软件生产数据，如果说数据是粮食，那么软件就是把粮食加工成食品的工具，两者完美结合，才能破茧成蝶，产生 BIM 产业的饕餮盛宴。

## 4.3　HIM 实现兼容性和互操作性

结构主义人类学的基本主张：我们所了解的外部世界是通过意识领悟到的，我们观察到的现象具有我们所赋予他们的特征，这是由我们意识的操作方式和人类大脑整理、解释刺激物的方式所决定的。这种整理过程一个十分重要的特征是，我们可以把环绕在我们周围的时间和空间连续的切割成片断，这样我们就能首先把环境看作是由大量可归属到某一个名目中去的个别事物组成的，把时间看成是孤立时间的序列组成的。同理，作为人，当我们制造人工产品时（各种各样的人造物），例如筹备典礼或书写以往的历史，我们就会模仿我们对自然的理解；于是，和我们认定自然事物是分裂和有序的一样，我们分割和排列了文化产物（Leach，1970）。这一段是埃蒙德·利奇对列维斯特劳斯结构主义人类学的简洁释义。

提到未来建造的样子，不同的人可能会有不同的预见。但是谈到未来的建造过程，所有的人都会认同同样的发展趋势：自动化、智能化。从 Grasshopper 的关键图我们发现：每个阶段的每项任务内容都是有着数据支撑且逻辑性很强的组织。如果我们把所有过程放在一张图中，把每个对象间不同过程中的逻辑线用不同颜色同时表示，我们会看到，我们整个建造过程就是一张网（图 4-5）。网中的每个节点就是参与建造的每个任务。

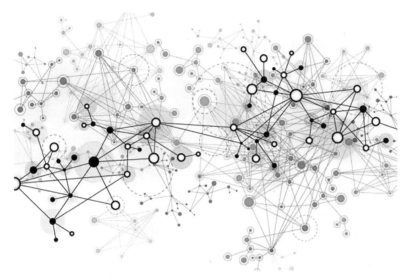

图 4-5　未来的建造过程

拓扑，是将各种物体的位置表示成抽象位置。在网络中，拓扑形象地描述了网络的安排和配置，包括各种节点和节点的相互关系。拓扑不关心事物的细节也不在乎什么相互的比例关系，只将讨论范围内的事物之间的相互关系表示出来，将这些事物之间的关系通过图表示出来。

HIM 作为未来建造的拓扑图的数学表达式如图 4-6 所示，左边图中的任何一个建造节点，均可通过右边的 HIM 有序分裂（详见图 4-7）。

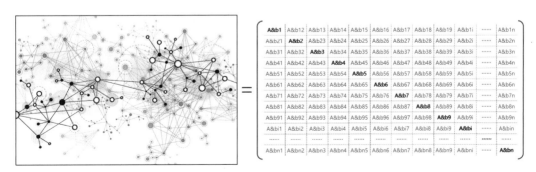

图 4-6　拓扑的 HIM 表达

图 4-7 复杂建造过程 HIM 解构

由此，形成基于 HIM 网络操作系统的智慧建造模式（图 4-8）。

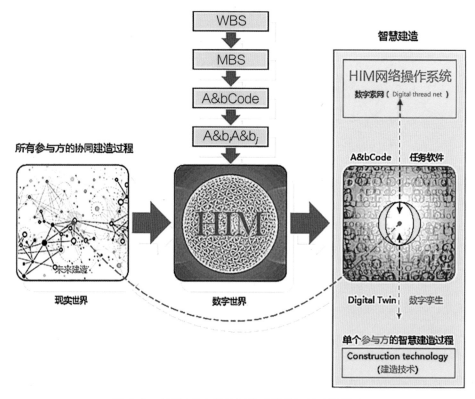

图 4-8 基于 HIM 网络操作系统的智慧建造模式

A&bCode 相当于网络中的一个固定 IP 地址，它能提供稳定的逻辑关系，从而使得工程逻辑变成数据交换逻辑预先植入到 HIM 建立的网络交换的规则，使得传输的数据从出发端可以通过 A&bCode 直达需求端，且信息可精准定制（图 4-9）。

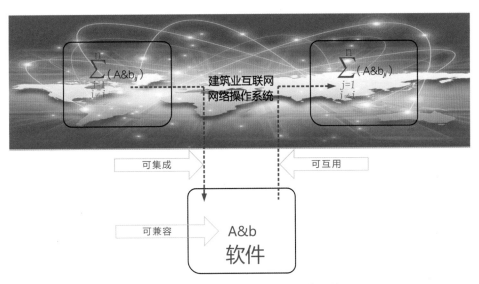

图 4-9 端对端的建筑业网络操作系统

# 第 5 章 A&bCode 与 BIM

## 5.1 美国 BIM 标准体系与中国 BIM 标准体系

美国 BIM 标准体系如图 5-1 所示。

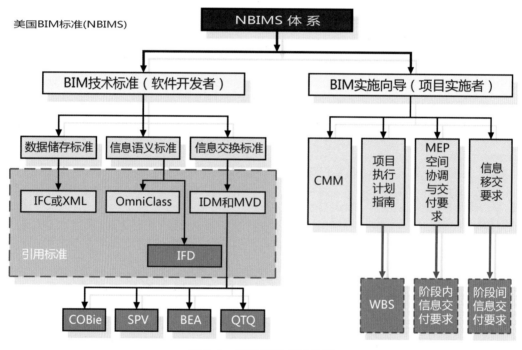

图 5-1 美国 BIM 标准体系

中国 P-BIM 标准体系如图 5-2 所示。

从图 5-1 与图 5-2 比较可见，美国 BIM 标准体系和中国 P-BIM 标准体系的区别仅在于主导 BIM 技术应用的决策者的不同认知。不同的认知产生不同的 BIM 实施方式，中国国家标准《建筑信息模型应用统一标准》中包含了两种不同的 BIM 实施方式（图 5-3）。

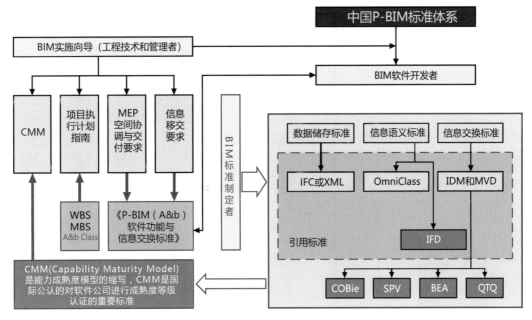

图 5-2 中国 P-BIM 标准体系

从图 5-3 可见，国际主流 BIM 标准体系采用了以 OmniClass 为代表的建筑业信息分类标准实施 BIM，而 P-BIM 标准体系则采用了以 A&bCode 为代表的建筑业信息分解标准实施 BIM。因此，P-BIM 标准体系兼容国际主流 BIM 标准体系。

## 5.2 A&bCode 与 OmniClass

将 A&bCode 编码方式与 OmniClass 的 15 张表对比，OmniClass 15 张表的内容可以分别对应于 A&bCode 编码中（图 5-4）。

## 5.3 A&bCode 与 IFC、NBIMS 信息交换架构

众所周知，IFC 大纲（总体框架）如图 5-5 所示，是分层和模块化。

整体分为四个层次，从下到上分别是资源层、信息核心层（信息框架层）、信息共享层和领域层。每个层次都包含一些信息描述模块，并且遵守一个原则：每个层次只能引用同层次和下层的信息资源，而不能引用上层资源。这样，上层资源变动时，下层资源不受影响，保证信息描述的稳定性，每个层次包含的内容：

1) 资源层描述标准中用到的基本信息，如人员信息、文档信息、几何拓扑信息等。这些基本信息不针对建筑工程与设备管理，仅仅是无整体结构的分散信息，它们将作为信

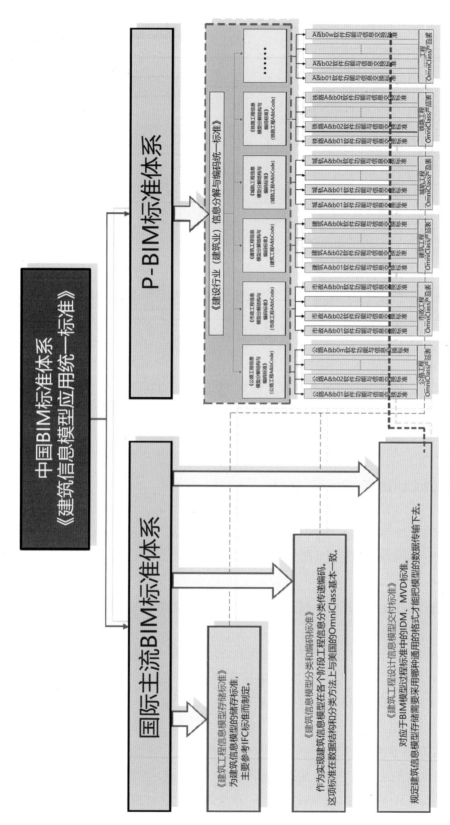

图 5-3 中国 BIM 标准体系

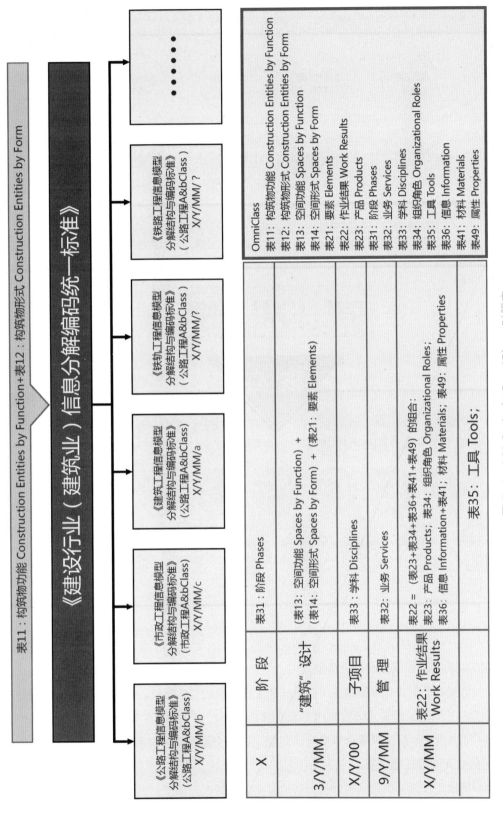

图5-4 A&bCode与OmniClass对照表

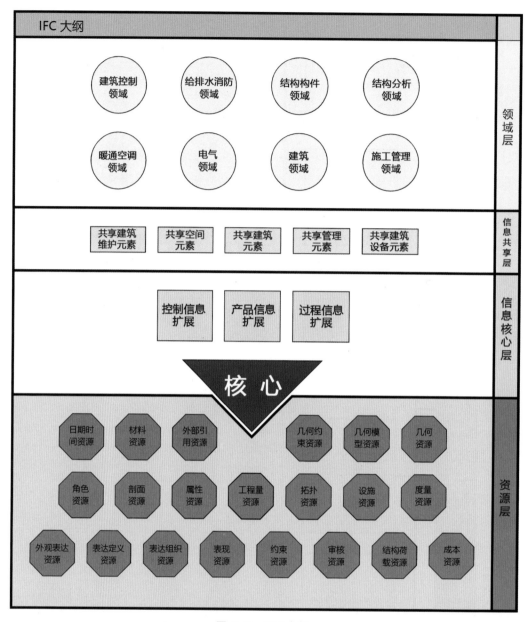

图 5-5 IFC 大纲

息描述的基础，应用于整个信息模型。

2）信息核心层（信息框架层）描述建筑工程信息的整体框架，它将资源层的信息用一个有规则的框架组织起来，使他们相互联系和连接，组成一个整体，真实反映现实世界的结构。

3）信息共享层解决领域信息交互的问题。在这个层次，各个系统细化明确组成元素，并保证组成元素不缺项遗漏。

4）领域层深入各个领域的内部，形成各个领域的专题信息。

每个层次内包含了若干模块。

IFC 建筑设计数据具有很强的特殊性。在 IFC 模式的 IFC2x2final 版本中定义了 623 个实体、110 个具体类型、159 个列举类型、42 个选择类型，而且各个实体之间存在着纷繁复杂的关系。正是这些实体的实例及其之间的关系，构成了完整的 IFC 建筑模型。

对于一般应用开发人员不需要了解 IFC 标准的全部内容，在清楚整体框架和核心结构的情况下，只需了解对应部分即可。

在 AEC/FM 行业内，IFC 相关的科研和实际应用项目越来越多，在这样的环境下，人们发现通过文件进行信息传输的方式，不利于行业的发展。因为在应用中，人们往往只需要文件内的部分信息，而通过文件传输只能传输整个模型，大大增加了系统开销，同时数据也不易用。因而，针对 IFC 局部模型数据的数据共享系统便应运而生。此后，局部模型数据交换一直都是 IFC 标准的相关研究中的热点问题，但是一直以来并未形成一个通用的、灵活的并具有较强扩展性的方法来描述 IFC 局部模型数据查询信息及实现 IFC 局部模型数据交换控制。

由于 IFC 标准应用的广泛性，会有各种各样的查询描述方法（或者说客户端程序）提供给用户用以描述所需要的局部模型数据。例如使用图像化的用户界面，允许设置某些特殊的值来选取局部模型，如：门高不超过 200cm 的门对象。同时，当不同种类的客户端程序连接到 IFC 局部模型服务器时，他们可能对局部模型的选择有多种不同的需要。例如：客户端程序 A 需要的是特定的建筑元素对象和相关的几何对象；而客户端程序 B 需要的是选择同样的建筑元素对象和其相关的造价信息对象。因此，IFC 局部模型共享服务器必须能够适用于上述不同的查询情况，给出相应的局部模型数据。

IFC 标准推广者一直在寻求一种可编程的查询语言来支持局部模型数据的操纵，既能满足客户终端对各式各样描述方法和内容的查询，又能避免用户交互上的困难。只有使用一种与程序无关的局部模型查询语言，才能避免将查询描述的处理算法直接编写在不同的程序里，从而保证系统的简洁性、灵活性和可扩展性。然而，建设项目全生命期过程中的局部模型需求千变万化，局部模型需求难以确定，这种查询语言无法编写出来。

工程项目的领域层仅包含实体和领域服务，实体为"建物"，包含设计、施工流程的不同"建物"方式，如图 5-5 中的"暖通空调领域"、"电气领域"、"建筑领域"；领域服务为"建物"管理行为，如"施工管理领域"、"物业管理领域"。

建筑工程项目是一项复杂的、综合的多专业活动，专业活动即为领域实体；管理领域实体的行为即为领域服务。也就是说图 5-5 中的第四层"领域层"应包含项目全生命期所有独立的"活动或商业"（图 1-4，美国 NBIMS 数据交换层次架构图）。

没有哪个软件开发商能够独立提供覆盖建筑物全生命期的应用系统，也没有哪个工程是只使用一家软件公司的软件产品就能完成的。因此，解决信息交换和共享问题对建设行业而言非常重要。目前的建筑软件只是涉及建筑工程全生命期的某个阶段中某个专业的领域应用。IFC 标准领域层（IFC-Domain Layer）作为 IFC 体系架构的顶层，定义了面向各个专业领域的实体类型。这些实体面向各个专业领域都具有特定的概念。为使 IFC 满足项目全生命期的信息交换要求，需要定义领域层的所有 A&b，即 A&bCode。

领域模型是对领域内概念类或现实世界中对象的可视化表示，领域模型也称为概念模型、领域对象模型和分析对象模型，专用于解释业务领域中重要的事物和产品。领域层是三层数据库结构中的概念模式（图 5-6）。

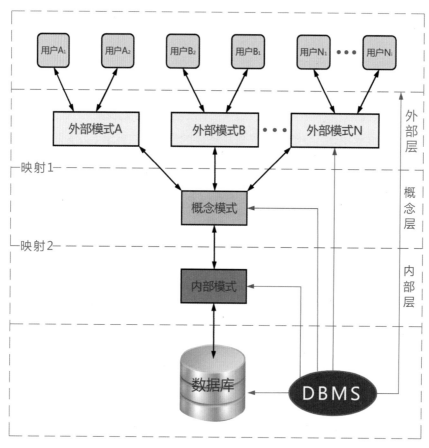

图 5-6　三层数据库结构

没有领域层的明确定义，IFC 总体框架的美好理想也就无法落地。IT 人员纵有十八般武艺也难以满足工程技术和管理人员千变万化的需求，这是 IT 之外的专业人员和管理者的任务，没有专业人员主导领域层的领域实体和领域服务分析，IFC 标准及 buildingSMART 和美国 BIM 标准的交互理论就难以落地为项目工程师所应用。

只有对 IFC 领域层进行深入分析，列出建筑工程全生命期所有 A&b，IT 工作者按照这些需求才有可能实现局部模型交互需求。A&bCode 的推出，填补了 IFC 架构领域层的空白，完善了 IFC 架构。领域层需要建立领域模型，建立领域模型的第一步就是需要识别出实体、值对象与领域服务。

- 实体：实体是领域中需要唯一标识的领域概念。通常在业务中，有一类对象需要唯一标识与区分，并持续对它进行跟踪，这样的对象我们认为是实体。这里的唯一标识通常指的是业务上的唯一标识，比如订单号、雇员工号等信息，而不是数据库中因为技术需要存储的 int 自增 id 或 Guid 列；实体只保留必要的属性与行为。比如对于一个客户实体，应该保留客户的基本信息，如：姓名、性别、年纪等；但像国家、省、城市、街道等信息联合表示客户完整的从属概念，这种完整的从属概念应该迁移到其他实体或值对象上，这样有助于对客户实体的理解和维护，并明确清晰的对象职责。

- 值对象：值对象是领域中不需要唯一标识的领域概念，通常在业务中，我们不需要区分对象具体是哪一个，而只关心对象是什么状态，这样的对象我们认为是值对象；如果两个对象所有状态都一样，我们就认为是同一个值对象，比如地址信息、订单状态信息等；值对象是只读的，具有不变性，不能直接修改，但可以被替换。

- 领域服务：某些业务行为既不能定义为某个实体，也不能归于值对象时，可以把它们归于领域服务这种概念。领域服务本质上就是一些操作，不包含状态，通常用于协调多个实体。领域服务可以直接被读取于应用层，这样可以有效地保护领域模型。

领域/应用层为 IFC 架构的最高层级，提供了营建和设施管理领域所需要的概念模型。目前 IFC 所定义的领域模型（Domain Models）包含建筑（Architecture）、设施管理（Facility Management）、估价（Cost Estimating）以及机电设备（HVAC），这种定义方式充其量也仅仅是设计阶段的部分领域模型，A&bCode 是基于建设工程的 WBS 和 MBS，是工程全生命期的所有领域模型。

IFC 之架构遵循阶梯原则 (Ladder principle)：每一个层级的类别可参照 (reference) 同一层级或较低层级的其他类别，但不能参照较高层级的类别。由此可见，位于领域层的 A&b 领域模型可以调用本层及下层的类别。亦即，制定《A&b 软件功能与信息交换标准》时，根据各种实际条件，可以完全利用 IFC 标准、部分应用 IFC 标准和完全不用 IFC 标准。这不仅弥补了不完善的建筑工程 IFC 标准，而且对暂时还没有 IFC 标准的基础设施工程 BIM 应用也极为重要。

因此，《A&bCode 编码体系标准》是 BIM 技术创新和落地实施的基础标准。

IFC 是中间数据格式，项目全生命期的所有软件都基于 IFC 进行交换，A&bCode 描述了 IFC 领域层的所有软件概念模式的集合，没有 A&bCode，即使有了 IFC 领域层的概念模式也无法实现外模式。美国 BIM 标准信息交换架构就是为不同的 A&bCode 制定外模式（图1-4），因此，缺乏 A&bCode，美国 BIM 标准也难以落地。

将美国 BIM 标准、IFC 大纲（图 5-5）、数据库三层结构（图 5-6）及 A&bCode 关系联接示意如图 5-7 所示。

将信息交换架构图 1-4 与图 5-7 关联得到图 5-8。

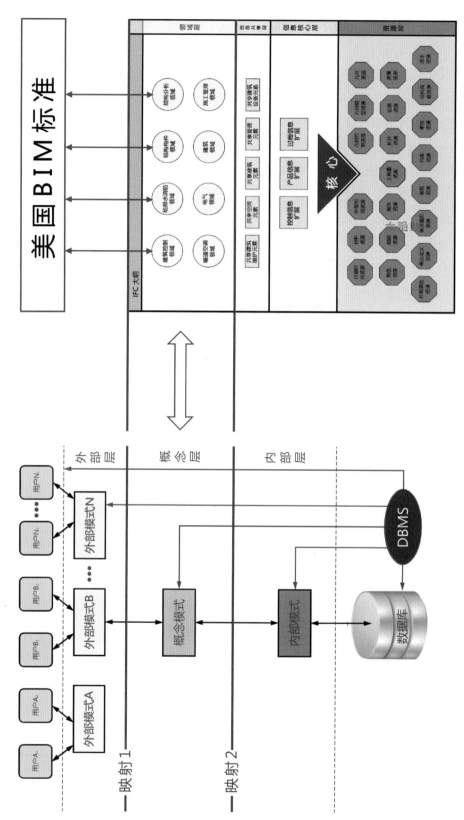

图 5-7 美国 BIM 标准的主要工作

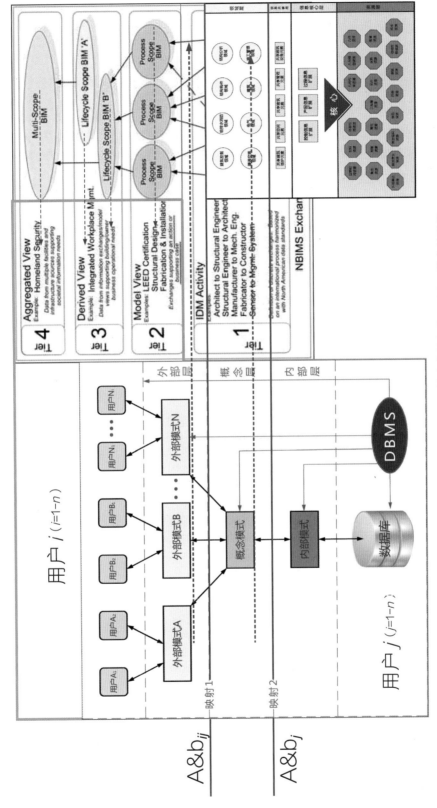

图 5-8 A&bCode、美国 BIM 标准信息交换架构、IFC、用户关系图

第5章 A&bCode与BIM 063

## 5.4　A&bCode 与 IDM/MVD

　　基于 A&bCode 的 HIM 与美国 BIM 标准信息交换架构、IFC、用户关系如图 5-9 所示。

　　图 5-9 将 IDM 直接以不同角色采用的 A&bCode 软件交换对应的 MVD 表达，即 IDM 与 MVD 完全映射。$A\&b_{ij}$ 与 IDM/MVD 概念相同。

　　A&bCode 细化了 IFC 的领域层和美国 BIM 标准信息交换架构中的应用层（图 5-10）。

　　我国国家标准《建筑信息模型应用统一标准》GB/T 51212—2016 中推荐的 P-BIM 标准与美国 NBIMS 实施 BIM 的区别列于图 5-11。

　　由图 5-11 可见，当 P-BIM 标准中的各应用软件间数据不交换时，即为图 5-11 右侧的 NBIMS 标准模式，但协同软件变成了 MVD 的建模软件。

　　由中国 BIM 发展联盟组织编写的系列标准《××P-BIM 软件功能与信息交换标准》已经完成建筑工程设计阶段的部分标准（图 5-12）。

　　由图 5-9 可见，基于 HIM 的 BIM 实施方式将图 5-13 所示的人类语言交换方式的 IDM，通过 A&bCode 直接改用计算机不同软件间的互操作（图 5-14）。这种方法就是以终定始：直接编制 MVD 标准，软件公司完成接口程序实现 BIM。

　　中国 BIM 发展联盟基于图 5-14 A&b 兼容软件互操作（基于 HIM 的建筑业互联网平台）实验室已经在深圳大学建成使用（图 5-15）。

　　基于 WBS 和 MBS 的建筑业信息分类 A&bCode 与工程实践紧密结合（贯穿建设工程全生命期）、与项目参与各方从业人员密切相关（以人为本），使 BIM 实施与工程任务一体化。通过开发分部分项工程（A&b）各节点用软件，以终定始，确定合约及设计阶段数据交换标准，是一种以简单需求为导向的 BIM 实施方式。为我国 BIM 创新创业（而不是建模）提供了巨大的发展空间。建筑信息模型分解结构确定的信息分解方法对我国 BIM 技术创新发展具有重要意义。

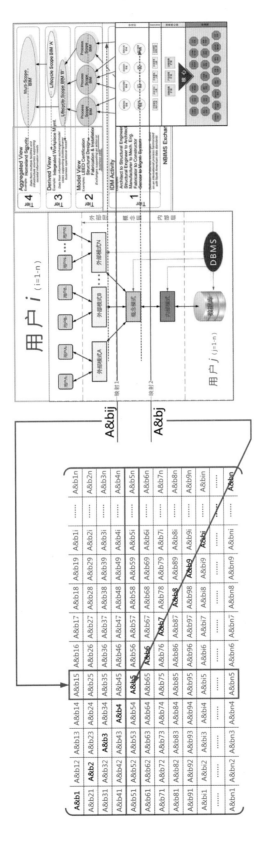

图 5-9 基于 A&bCode 的 HIM

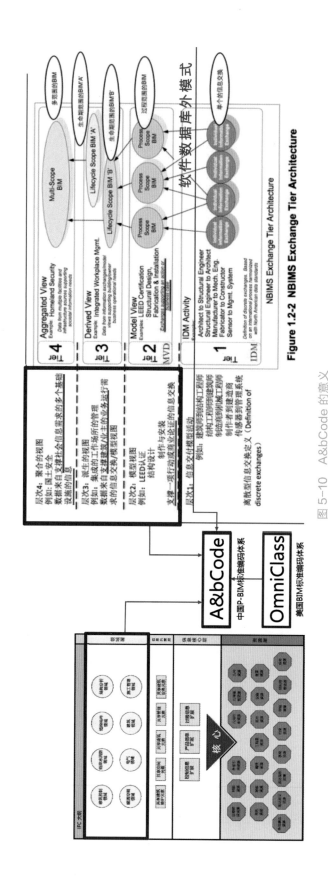

图 5-10 A&bCode 的意义

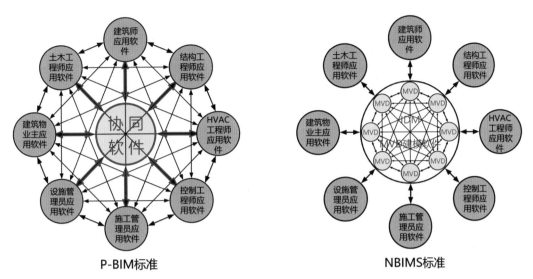

图 5-11　P-BIM 标准与美国 BIM 标准

图 5-12　《××P-BIM 软件功能与信息交换标准》

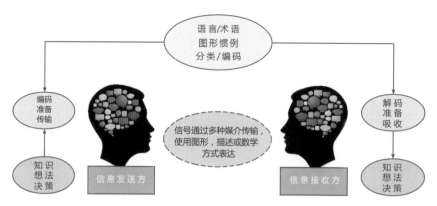

图 5-13 IDM 人类语言交换标准

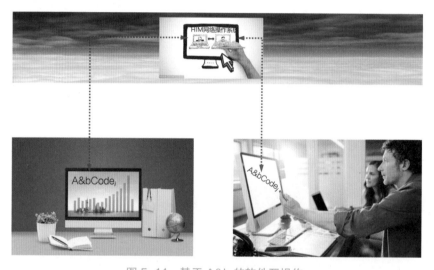

图 5-14 基于 A&b 的软件互操作

图 5-15 基于 HIM 建筑业互联网平台

# 致　谢

　　我从 2012 初年开始研究 BIM，至今整七年，工作与生活中的绝大部分时间都花在这上面。特别感谢中国 BIM 发展联盟许杰峰、朱雷、金新阳、楼跃清、金睿、伍军、黄琨、刘洪舟、李邵建、龚剑、李云贵、张建平、徐建中、左江、王晓军、程志军、何关培、甘嘉恒、毛志兵、王丹、谢卫、高承勇、王家远、李东彬、王荣、张淼、任菲菲、张峥对我的全力帮助与支持。

　　今天我带领的 A&bCode 研究组的这项研究成果得以出版，七年来指导我、帮助我、关心我的人与事至今历历在目。

　　2012 年 4 月，北京理正软件设计研究院有限公司为我设计了 P-BIM 的概念图，后来成为中国 BIM 发展联盟 LOGO。我及同事们一直沿着这个目标努力。

　　2012 年 6 月，我参加住房和城乡建设部 BIM 考察团到美国访问，在甘嘉恒博士全程陪同下，使我认识了 BIM 的几个重要观点。

1. 美国 buildingSMART alliance Deke Smith 先生：BIM 发展和实施过程中可能的最大风险是"overselling（过度吹嘘）"。

2. 佛罗里达迪斯尼 Imagineering（想象工程）团队 Mark A. Kohl 先生：What you get out of BIM is a result of what you put into BIM（你从 BIM 中所得到的即是你放入 BIM 中的东西的结果）。

3. Autodesk 负责全球企业战略的副总裁 Jon Pittman 先生：我们是工具制造商，不是我们自己使世界变得更精彩，我们只是制造工具辅助设计师使世界变得更精彩。

4. 斯坦福大学 CIFE 中心主任 Martin Fischer 先生：我们要为下一代从业人员准备好他们愿意在这里（建筑业）工作的环境。

5. buildingSMART 国际总部主席 Patrick MacLeamy 先生：你们为什么那么热衷于 BIM？

6. 美国 HOK 公司的 VDC/BIM 团队：BIM 应该方便使用，不同软件产生的数据存放在 Excel 表里面。

带着这些观点和问题，2012 年以来我们一直致力于建立中国 P-BIM 标准体系。感谢这些年所有参与 P-BIM 标准研究的单位和个人及其成果。

感谢深圳大学为 P-BIM 软件建立了数据交换实验室（建筑互联网与 BIM 实验室）。

2017 年 11 月，buildingSMART 国际总部主席 Patrick MacLeamy 和首席执行官 Richard Petrie 到中国 BIM 发展联盟访问，深入交流了 openBIM 和 P-BIM。

2018年6月,在深圳与到访的澳大利亚Richard Choy教授对P-BIM体系深入讨论中,提出了建筑业信息分解编码体系A&bCode(原来定名为A&bClass)。

感谢本书插图汇编人员:张良、孙文成、刘巨龙、周炳松、郑锦州、蔡耀聪。
感谢中国BIM发展联盟高级研修班的所有老师和学员们。

黄强

2019年1月19日

# A&bCode 软件数据交换演示

2017年5月，由中国BIM发展联盟与深圳大学共同发起成立的中国BIM发展联盟深圳大学建筑互联网与BIM实验研究中心，正式运行使用。BIM实验研究中心围绕P-BIM实施方式，建设BIM应用示范研究基地，搭建国内外软件协同工作实验研究平台，实现多专业BIM技术协同工作。

以深圳大学BIM实验研究中心为平台，联盟组织建研科技股份有限公司、北京理正软件股份有限公司、香港图软亚洲有限公司北京代表处、天宝蒂必欧信息技术（上海）有限公司、欧特克软件（中国）有限公司、北京鸿业同行科技有限公司、杭州品茗安控信息技术股份有限公司、上海建筑科学研究院（集团）有限公司、中国建筑技术集团有限公司、建研地基基础工程有限责任公司、广东星层建筑科技股份有限公司、广东省建筑科学研究院集团股份有限公司、山东省建筑科学研究院、上海市基础工程集团有限公司、福建省建筑设计研究院、建研科工（北京）建筑工程技术有限公司、上海宾孚建设工程顾问有限公司、昆明市建筑设计研究院集团有限公司、深圳市北斗云信息技术有限公司等20多家企业开展联盟技术研发工作，针对各应用软件研发的标准接口已在联盟深圳大学BIM实验研究中心实现数据无缝对接，协同创新工作不断深入。

| 软件功能名称 | 规划设计 | 建筑设计 | 勘察设计 | 地基设计 | 结构机电碰撞检查 | 混凝土结构设计 | 给排水设计 | 暖通空调设计 | 电气设计 | 灌注桩合约 | 钢筋工程合约 | 混凝土工合约 | 脚手架工合约 | 地基试验检测 | 质监站项目管理 |
|---|---|---|---|---|---|---|---|---|---|---|---|---|---|---|---|
| 规划设计 | 规划设计2000a | 2000a3000a | 0 | 0 | 0 | 0 | 0 | 0 | 0 | 0 | 0 | 0 | 0 | 0 | 2000a9253a |
| 建筑设计 | 0 | 建筑设计3000a | 0 | 0 | 3000a3200b | 3000a3202a | 3000a3301a | 3000a3302a | 3000a3303a | 0 | 0 | 0 | 0 | 0 | 3000a9253a |
| 勘察设计 | 0 | 0 | 勘察设计3101a | 3010a3104a | 0 | 0 | 0 | 0 | 0 | 0 | 0 | 0 | 0 | 0 | 3101a9253a |
| 地基设计 | 0 | 3014a3000a | 3014a3101a | 地基设计3104a | 0 | 0 | 0 | 0 | 0 | 3104a4101a | 0 | 0 | 0 | 0 | 3104a9253a |
| 结构机电碰撞检查 | 0 | 3200a3000a | 0 | 0 | 结构机电碰撞检查3200b | 3200b3202a | 3200b3301a | 3200b3302a | 3200b3303a | 0 | 0 | 0 | 0 | 0 | 0 |
| 混凝土结构设计 | 0 | 3202a3000a | 0 | 0 | 3202a3200b | 混凝土结构设计3202a | 0 | 0 | 0 | 0 | 0 | 3202a4211a | 3202a4213a | 3202a4214a | 3202a9253a |
| 给排水设计 | 0 | 3301a3000a | 0 | 0 | 3301a3200b | 0 | 给排水设计3301a | 0 | 0 | 0 | 0 | 0 | 0 | 0 | 3301a9253a |
| 暖通空调设计 | 0 | 3302a3000a | 0 | 0 | 3302a3200b | 0 | 0 | 暖通空调设计3302a | 0 | 0 | 0 | 0 | 0 | 0 | 3302a9253a |
| 电气设计 | 0 | 3303a3000a | 0 | 0 | 3303a3200b | 0 | 0 | 0 | 电气设计3303a | 0 | 0 | 0 | 0 | 0 | 3303a9253a |
| 灌注桩合约 | 0 | 0 | 4101a3101a | 4101a3104a | 0 | 0 | 0 | 0 | 0 | 灌注桩合约4101a | 0 | 0 | 0 | 0 | 4101a9253a |
| 钢筋工程合约 | 0 | 0 | 0 | 0 | 0 | 4211a3202a | 0 | 0 | 0 | 0 | 钢筋工程合约4211a | 0 | 0 | 0 | 4211a9253a |
| 混凝土工合约 | 0 | 0 | 0 | 0 | 0 | 4213a3202a | 0 | 0 | 0 | 0 | 0 | 混凝土工合约4213a | 0 | 0 | 4213a9253a |
| 脚手架工合约 | 0 | 0 | 0 | 0 | 0 | 4214a3202a | 0 | 0 | 0 | 0 | 0 | 0 | 脚手架工合约4214a | 0 | 4214a9253a |
| 地基试验检测 | 0 | 0 | 0 | 9102a3104a | 0 | 0 | 0 | 0 | 0 | 0 | 0 | 0 | 0 | 地基试验检测9102a | 0 |
| 质监站项目管理 | 9253a2000a | 9253a3000a | 9253a3101a | 9253a3104a | 0 | 9253a3202a | 9253a3301a | 9253a3302a | 9253a3303a | 9253a4101a | 9253a4211a | 9253a4213a | 9253a4214a | 0 | 质监站项目管理9253a |
| 实验室具体软件名称 | 理正 | 图软欧特克 | 理正 | 建研科工 | 欧特克 | PKPM | 鸿业 | 鸿业 | 鸿业 | 地研地基上海基础 Tekla | 星辰 | 星辰 | 品茗 | 广东建科研 | 山东建科院 |

联盟深圳大学BIM实验研究中心实现数据无缝对接

中国 BIM 发展联盟特邀常务理事、美国斯坦福大学综合设施工程中心项目总监及兼职教授、bimSCORE 创始人及 CEO，甘嘉恒博士（Dr. Calvin Kam）在深圳大学 BIM 实验室对 A&b 兼容软件互操作（基于 HIM 的建筑业互联网平台）情况进行了全英文的详细描述讲解。

讲解视频（时长 17 分钟）可扫码观看。

若想下载请访问以下链接。
【百度网盘】https://pan.baidu.com/s/1kvNzt62wCAfViNl9G4C0aA 提取码：8r2h

本书由下列人员翻译：

序：吴露方

第 1 章：王照华

第 2 章：都浩、吴露方

第 3 章：张淼、王照华、徐霄枭和杨嵩桥

第 4 章：赵峰

第 5 章：蒲秋兴

致　谢：王照华

全文由吴露方、徐霄枭、王照华和蒲秋兴校对，并由吴露方进行统稿完成。

本文翻译水平有限，存在不少错误之处望提出批评改正，有不清楚之处请联系吴露方博士

联系方式：wulufang@xingshunit.com

This article is translated by:

Preface: WU Lufang

Chapter 1: WANG Zhaohua

Chapter 2: DU Hao, WU Lufang

Chapter 3: ZHANG Miao, WANG Zhaohua, XU Xiaoxiao and YANG Songqiao

Chapter 4: ZHAO Feng

Chapter 5: PU Qiuxing

Acknowledgements: WANG Zhaohua

The full text was proofread by WU Lufang, XU Xiaoxiao, WANG Zhaohua and PU Qiuxing, and it was finally completed by WU Lufang.

The translators and interpreters' level of this article is limited, we are sorry that there are many mistakes, and it is hoped that criticism and correction will be made. If you have any questions, please contact Dr. Wu Lufang.

E-mail: wulufang@xingshunit.com

# Preface

*Tao Te Ching* says: "Its largest square doth yet no corner show; a semblance great, the shadow of a shade", which is the highest state of Laozi "Tao". Everything in the world is often not rigidly adhere to their own genes, showing the "meteorological" form and pattern, information technology is also so, seemingly obscure and abstruse, but combined with the reality work, it is also simple. This is a universal and dialectical materialist world view.

Interoperability between different applications ( functional docking, data transfer, model exchange, information collaboration ) can be narrowly defined as "one application sends data to another". In a broad sense, it is possible that different applications can mutually retrieve and effectively use each other's data. But in any case, the object of communication between different applications can only be data, even if it is a tangible three-dimensional component shape, it is expressed in the form of data in applications. When we turn off the computer, the application will fold all the 3D models and collapse to the corresponding space, which will save the data to a database.

Therefore, Interoperability is data Interoperability between applications.

The objective of NBIMS is the interoperability of BIM applications.

The Industrial Internet effort will bring industrial control systems online to form large end-to-end systems, connecting them with people, and fully integrating them with enterprise systems, business processes and analytics solutions. These end-to-end systems are referred to as Industrial Internet Systems ( IISs ). Within these IISs, operational sensor data and the interactions of people with the systems may be combined with organizational or public information for advanced analytics and other advanced data processing ( e.g., rule-based policies and decision systems ). The result of such analytics and processing will in turn enable significant advances in optimizing decision-making, operation and collaboration among a large number of increasingly autonomous control systems.

Industrial internet systems are composed of many components provided by different vendors and organizations. For these parts to be put together successfully, they must have the following properties:

Integration - refers to the ability of a component to communicate with other components through various signals and protocols.

Interoperability - the ability to understand information based on a common conceptual framework and unified context.

Compatibility - the ability to interact with other components as required by reengineering and to meet the requirements of the interactor.

Obviously, Compatibility is the best way to achieve industrial Internet. Compatibility depends on and contributes to interoperability and integration.

The objective of China's P-BIM standard system is the application compatibility.

Great Tao is simple. When the system is analyzed, the key is not only understanding the whole. It is still needed to have a thorough understanding of things' internal structure and composition, various components of the integration and coordination between the relationships. In order to accomplish the task, the most important job is to be independent from each other, totally exhausted, level of granularity appropriate breakdown, in order to achieve the skilled and magical craftsmanship, and be a complete master. The process of breakdown is a top-down process, while the process of classification and abstraction happens to be a bottom-up process. The purpose of breakdown is from the whole to the individual. The purpose of classification is from the individual to the whole, through the classification and abstraction to achieve the abstract category of unified decision-making and action. Therefore, the classification that does not break down the information of the Architecture, Engineering & Construction ( AEC ) industry is necessarily a huge and difficult to control. The combination of top-down breakdown of construction information with appropriate level granularity and bottom-up classification and unified decision-making lies in Summary Work Breakdown Structure ( SWBS ) of the AEC sub-industry system.

Enlightenment comes from nature. To make different applications in the same system (subsystem) and meet the requirement of compatibility, it is needed to outline work breakdown system (task), and set up coding system of different function subsystem on each task in the system. Subsequently, taking advantage of the logical relationship between coding subsystem can realize the two-way interaction between different subsystem applications folder. At the same time, the compatibility of applications can be realized by defining classified interactive content and exchange data format standard for exchange folders.

In 1993, ISO/TR 14177 technical report titled *Classification of Information in the Construction Industry* embarked the construction industry on the journey of information classification. 26 years later, we are going to creatively launch *Breakdown and Coding of Construction Industry Information A&bCode*, Which is a coding system for the construction industry, aiming to satisfy the informatization developments of construction Industry.

In the objcect-oriented method, concept of essential characteristics of things abstracted from entities with common attributes is "class"; in the systems engineering-oriented method, elements in the systems are regarded as "alive" subjects with its own purposes, initiative and enthusiasm, and it is called "code". We can restructure the function of construction industry application system by means of A&bCode, so as to modularize the AEC industry application tools and generate AEC industry applications in a smaller, simpler, more professional and more delicate way; Establish the AEC industry information network exchange system based on A&bCode in order to further develop and facilitate the AEC industry internet network system towards four primary objectives of "minimal network architecture, minimal internet trade pattern, maximum network security and privacy protection follows GDPR".

Similar to internet network architecture, several rules are suggested when it comes to establishing AEC industry internet network structure: the ACE industry network, equipment, and system data exchange interface code A&bCode should be the equivalent of the fixed IP address of internet public network; the A&bCode series of standards for P-BIM applications function and information exchange should be the equivalent of the TCP/IP of Internet Protocol Suite; the HIM network operating system of A&bCode-based AEC Industry internet should be the equivalent of the Internet Domain Name System (DNS).

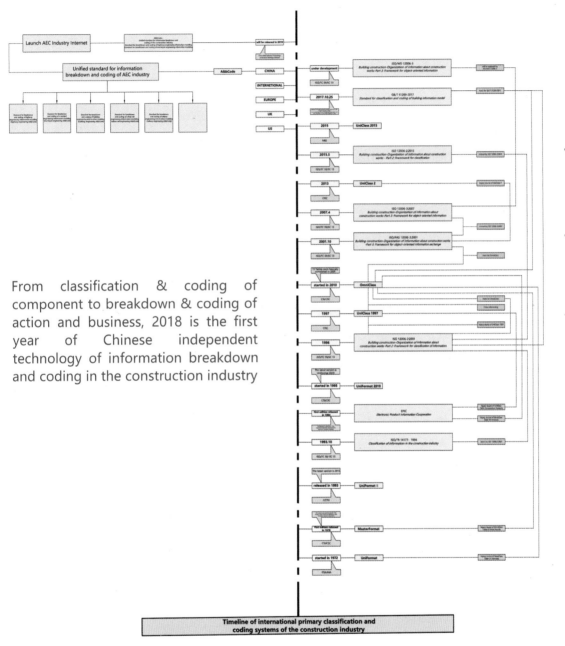

Figure 1　The history of information classification for the construction industry

# Contents

Preface

| | | |
|---|---|---|
| **Chapter 1** | **Interoperability Characteristics of BIM** | **085** |
| 1.1 | Text-based Systems Engineering & Model-based Systems Engineering | 086 |
| 1.2 | P2P | 087 |
| 1.3 | NBIMS Exchange Tier Architecture | 088 |
| 1.4 | From "Point to Point" to "One to One" and back to "Point to Point" | 089 |
| 1.5 | OpenBIM | 093 |

| | | |
|---|---|---|
| **Chapter 2** | **Coding System for Classification of Information in the Construction Industry: OmniClass** | **094** |
| 2.1 | Hierarchical Classification Method and Faceted Classification Method | 094 |
| 2.2 | MasterFormat and UniFormat | 095 |
| 2.3 | Coding System for Classification of Information in the Construction Industry: Omniclass and Uniclass | 098 |
| 2.4 | Problems of BIM Classification Coding | 105 |

| | | |
|---|---|---|
| **Chapter 3** | **The A&bCode System for Breakdown and Coding of Information in the Construction Industry** | **108** |
| 3.1 | Breakdown and Classification | 108 |
| 3.2 | Work Breakdown Structure (WBS) | 114 |
| 3.3 | Pattern | 120 |
| 3.4 | WBS in the AEC Industry | 123 |
| 3.5 | MBS | 125 |

3.6 Distributed Functional Modeling Application
   (P-BIM Functional Application) ································ 126
3.7 P-BIM Functional Application Information Exchange Standard ········· 131
3.8 Construction Information Breakdown Coding System ················ 132
3.9 A&bCode Core's Ideas and Significances ························ 137
3.10 A&bCode's Demonstration of Coding Standards (Highway Engineering) ··· 139

## Chapter 4  HIM Interoperability Based on A&bCode ···················· 144

4.1 Digital Thread Network ·································· 144
4.2 HIM Digital Thread Network Based on A&bCode ···················· 146
4.3 HIM for Compatibility and Interoperability ························ 150

## Chapter 5  BIM with A&bCode ································· 154

5.1 BIM Standard Systems of the United States and China ················ 154
5.2 A&bCode and OmniClass ·································· 155
5.3 Information Exchange Architecture between A&bCode and IFC,NBIMS ··· 155
5.4 A&bCode and IDM/MVD ··································· 166

Acknowledgements ·········································· 171
Introduction of Advanced Training on BIM Application & Industry
  Collaborative Innovation ······································ 178
A&bCode Demonstration of Software Data Exchange ··················· 186
Figure: A&bCode and AEC Industry Internet ·························· insert

# Chapter 1  Interoperability Characteristics of BIM

The first section of NBIMS-US Version 1 reads:

Application interoperability, referring to the seamless exchange of information between different applications with independent internal data architecture, can be realized by mapping components of the internal data architecture of these programs to a universal data model, and vice versa. With an open universal data model, any application can be mapped and thus develops an interoperable relationship with any other one. Interoperability significantly reduces cost by eliminating the integration process of one application ( of different versions ) with another ( of different versions ) .

Reference standards of NBIMS-US provide basic definitions of entities, properties, relations and classifications independent of the computer, which are critical to conveying the abundant words of the construction industry. Reference standards selected by the NBIMS-US Committee are all international standards that have already obtained an ideal effect in terms of the capability of sharing complex design and the content of a construction project. The first section of NBIMS-US Version 1 includes three candidates of reference standards: IFC of IAI, OmniClass of CSI and IFD of CSI.

IFC data model, consisting of definitions, rules and protocols, defines the data set which describes the lifecycle of the building in a unique way. These definitions allow the application developer to write the IFC interface into the applications so that it can communicate and share information of the same format with other applications regardless of the internal data architecture. Applications with IFC interface can also exchange and share information with each other.

OmniClass or OCCS is a multi-list classification system for the basic construction industry, including the most common classification methods and important information of NBIMS-US with different forms of organizations, supplied in electronic edition and hard copy. OCCS can be used to prepare various types of project information, exchange information, cost, specification and other information produced during the lifecycle of the building.

buildingSMART Data Dictionary ( bSDD ) is a dictionary of architecture terms which must achieve an identical result under various contexts of language. Design of NBIMS-US depends on consistent term and classification ( based on OmniClass ) to support the interoperability of the model. The input item of OmniClass list can be used repeatedly once being defined in IFD Library. It can bring reliable automatic communication between applications, namely the major objective of NBIMS-US.

The common objective of BIM recommended by buildingSMART openBIM and NBIMS-US is interoperability ( Figure 1-1 ) .

Figure 1-1  The objective of BIM: Interoperability.

## 1.1  Text-based Systems Engineering & Model-based Systems Engineering

Traditional Systems Engineering is centered on the document which is text-based ( Figure 1-2 ) . In other words, Traditional Systems Engineering is the Text-based Systems Engineering ( TSE ) .

*The Vision of Systems Engineering 2020* defines Model-based Systems Engineering ( Figure 1-3 ) : Systems Engineering based on the system architecture model is the formalized application of modeling in order to support such activities as system requirements, design, analysis, verification and confirmation, which start from the conceptual phase and continue existing throughout design development as well as all subsequent phases of the lifecycle.

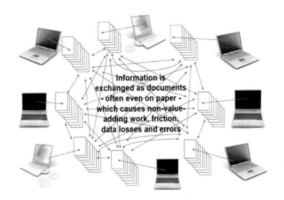

Figure 1-2　Text-based Systems Engineering ( TSE )

Figure 1-3　Systems Engineering based on the (system architecture) model

## 1.2　P2P

Peer to peer network (P2P), or peer to peer computer network, is a distributed application architecture for assigning tasks and workloads between peers. It is also a type of networking or network form developed by the peer to peer computing model in the application layer.

Simply, users can interact without obstacle through the internet based on P2P, which connects users directly. P2P eliminates intermediaries in order to make it easier to communicate, share, and interact more directly. The other critical attribute of P2P is the change from current centralization to un-centralized and give the rights back to end users.

P2P network is a concept of network architecture. One essential difference between P2P and dominant Client/Server (C/S) and Browser/Server (B/S) architecture (that is WWW) is that there is no central peer (or central server) in the entire network architecture within P2P. In the P2P architecture, each peer corresponds to the functions of information, i.e., consumer, provider and communication. In terms of computing pattern, P2P provides each peer with equal status by breaking conventional C/S or B/S pattern. Each peer serves as a server to provide services for other peers, while obtaining the services from other peers.

Compared with C/S or B/S networks, P2P has the following advantages:

1) Content and resources can be shared in the central and edge region. On the contrary, content and resources are usually shared only in the central area of the C/S or B/S network.

2) P2P network is usually easy to be extended and it is more reliable than a single server. Under a high network usage, a single server will encounter bottleneck as a result of the failure of a single point.

3) P2P network will share a common server in order to integrate computing resources and carry out tasks, rather than relying on a single computer, such as a supercomputer.

4) Users can directly access shared resources on the peer computer. And peers within the network can also share documents directly on local memory instead of a central server.

## 1.3 NBIMS Exchange Tier Architecture

The tier architecture of NBIMS for data exchange is shown in Figure 1-4. The information exchange architecture consists of IDM Activity, Model View, Derived View and Aggregate View tier. In the Model View tier, IFC defines an information exchange mode of action or business through MVD. Classes, attribute, relationships, attribute sets, quantitative definitions and others in all IFC concepts are used in this subset, which represents application requirements specifications. IFC interface (intermediate file format) is implemented to satisfy the information exchange requirements, while a specific independently IFC version is important for the exchange between different tiers in order to realize the operation of information sets of classes, attributes, relationships, attributes sets, quantitative definitions and other subsets, intersections, unions and complements of IFC with a larger (multiple) range.

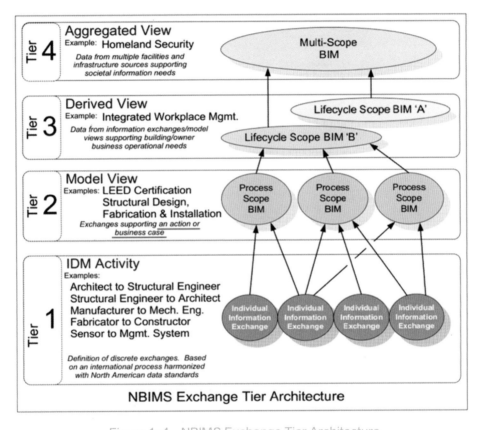

Figure 1-4　NBIMS Exchange Tier Architecture

## 1.4　From "Point to Point" to "One to One" and back to "Point to Point"

The conventional "point to point" (Text-based) text exchange approach is usually described in Figure 1-5 as a "one to one" (Model-based) exchange:

Actually, there is a serious misunderstanding in Figure 1-5. The left side is based on text exchange, which represents information exchange between different people. However, the right side denotes that different professionals should use applications to exchange with BIM database under a BIM working pattern. As a result, the object

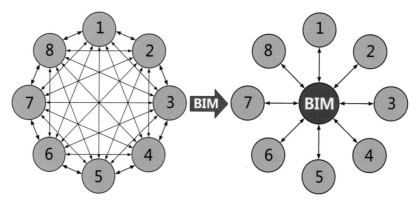

1. Architect
2. Structural Engineer
3. HVAC Engineer
4. Control Engineer
5. Construction Manager
6. Facility Manager
7. Owner
8. Civil Engineer

Figure 1-5  "Point to Point" and "One to One"

corresponding to the BIM on the right of Figure 1-5 is an application but not a specific person. More precisely, the object of BIM on the right of Figure 1-5 is application, and the text exchange on the left should be replaced by application exchange for matching, that is, Figure 1-6 should substitute for the left side of Figure 1-5.

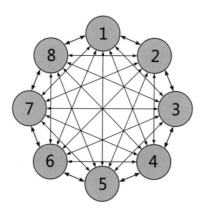

1. Applications for Architect
2. Applications for Structural Engineer
3. Applications for HVAC Engineer
4. Applications for Control Engineer
5. Applications for Construction Manager
6. Applications for Facility Manager
7. Applications for Owner
8. Applications for Civil Engineer

Figure 1-6  Point to Point information exchange between applications

There is no such thing as a free lunch. The nature of engineering information exchange will not be changed by simply drawing. It can be seen from Figure 1-4 that, the essence of information exchange among professionals in Figure 1-5 has not changed. It is a process of transforming text exchange (human language, IDM) on the left of Figure 1-5 into computer language (model view, MVD). The process is to compile IDM based on the original exchange content on the left of Figure 1-5, and then all IDM of applications are integrated to formulate the MVD. According to the NBIMS exchange tier architecture (Figure 1-4), the implementation roadmap of the BIM on the right side of Figure 1-5 is shown in Figure 1-7 below.

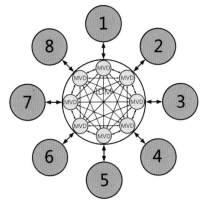

1  APPLICATIONS FOR ARCHITECT
2  APPLICATIONS FOR STRUCTURAL ENGINEER
3  APPLICATIONS FOR HVAC ENGINEER
4  APPLICATIONS FOR CONTROL ENGINEER
5  APPLICATIONS FOR CONSTRUCTION MANAGER
6  APPLICATIONS FOR FACILITY MANAGER
7  APPLICATIONS FOR OWNER
8  APPLICATIONS FOR CIVIL ENGINEER

Figure 1-7  Implementation roadmap of NBIMS-US

In summary, Figure 1-5 should be corrected to Figure 1-8.

As afore-mentioned, it can be realized that many problems in the current BIM implementation can be addressed through the application of P2P for data exchange between different applications. The left side of Figure 1-8 is a typical approach of P2P network exchange.

The critical problem of BIM implementation approach on the left side of Figure 1-8 is the insufficiency of collaboration of information exchange between different applications. To accomplish BIM in the approach of the left side of Figure 1-6, a generalized collaborative application (Figure 1-9) is indispensable because of the different phases

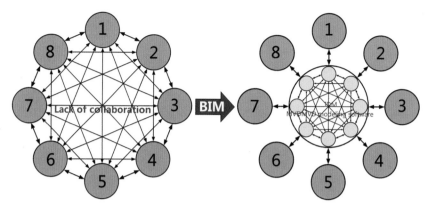

1 APPLICATIONS FOR ARCHITECT
2 APPLICATIONS FOR STRUCTURAL ENGINEER
3 APPLICATIONS FOR HVAC ENGINEER
4 APPLICATIONS FOR CONTROL ENGINEER
5 APPLICATIONS FOR CONSTRUCTION MANAGER
6 APPLICATIONS FOR FACILITY MANAGER
7 APPLICATIONS FOR OWNER
8 APPLICATIONS FOR CIVIL ENGINEER

Figure 1-8　Data exchange between appications from Point to Point to One to One (NBIMS-US)

of construction project with different collaboration content, for example, the primary collaborative content during the design phase is spatial collaboration among different professionals, but the coordination of progress and working faces between various construction methods in construction phase.

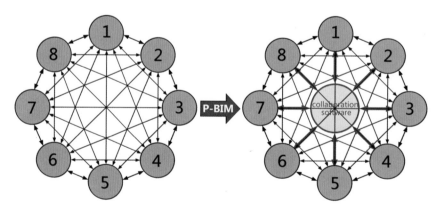

1 APPLICATIONS FOR ARCHITECT
2 APPLICATIONS FOR STRUCTURAL ENGINEER
3 APPLICATIONS FOR HVAC ENGINEER
4 APPLICATIONS FOR CONTROL ENGINEER
5 APPLICATIONS FOR CONSTRUCTION MANAGER
6 APPLICATIONS FOR FACILITY MANAGER
7 APPLICATIONS FOR OWNER
8 APPLICATIONS FOR CIVIL ENGINEER

Figure 1-9　Implementation roadmap of P-BIM

The right part of Figure 1-9 is still P2P network.

## 1.5 OpenBIM

OpenBIM is a universal approach to the collaborative design, realization, and operation of buildings based on open standards and workflows. In communication with the president and CEO of the International buildingSMART, the implementation objective of openBIM is the seamless exchange of information between different applications (as shown in Figure 1-10). It can be seen from Figure 1-10 that the data exchange of openBIM is an application approach directly realize the data delivery to receiver applications based on a single IDM. This approach is more like the left side of Figure 1-9 and P2P pattern.

International buildingSMART is trying to build an MVD library for different applications(as shown in Figure 1-11).

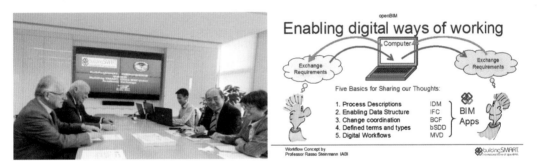

Figure 1-10  Data flow mode of openBIM

Figure 1-11  Assumption of BIM realization mode of buildingSMART

# Chapter 2  Coding System for Classification of Information in the Construction Industry: OmniClass

The purpose of establishing coding system is to scientifically and effectively manage the whole process of construction projects and standardize the behavior of project participants. Specifically, it is conducive to the control of the contents of the work in each phase of the project by the project construction units, such as the management and control of the total cost of the project, the implementation of value engineering research. It provides information exchange tools for project members, especially for information communication among development units, design units and construction units. It can effectively convey information while eliminating misunderstandings. In addition, the classification and coding of engineering information is the basis for data collection, summary, collation and analysis of engineering projects, and provides a guarantee for the use of accurate and valuable information for future projects.

## 2.1  Hierarchical Classification Method and Faceted Classification Method

Hierarchical classification method is to divide objects into several levels according to selected attributes (or features), and each level is divided into several categories. The same-level categories of the unified branch form a parallel relationship, and the different-level categories form a subordinate relationship. Categories at the same level do not repeat and overlap with each other.

Advantages: Firstly, the operation is more humane and conforms to the traditional application habits. It is suitable for manual processing and easy for computer processing. Second, the ability to expansion is good. Thirdly, the retrieval method of codes is very efficient. Fourthly, the data can be managed at different levels with good hierarchy, which can better reflect the logical relationship between categories.

Disadvantages: (1) The ability to reveal specific features of subjects or things is poor, often unable to meet the needs of accurate classification, cannot fully reflect the problems of the current large number of small classifications. (2) Classification table has certain solidification, it is not convenient to change at any time according to needs, and it is not suitable for multi-angle information retrieval. (3) It is impossible to automatically generate new classes according to the development of modern science and to keep pace with the development of science. (4) Large-scale classification tables generally have detailed categories and are fairly long, which has higher requirements for the management of classification tables. (5) The flexibility of classification structure is poor.

Surface classification method, also known as parallel classification method, is to divide the whole set of objects into surfaces without subordinate relationship according to their inherent attributes or characteristics, each of which contains a group of categories. We can combine one category in one surface with one category in another surface to form a composite category.

Advantages: The main advantage is that the classification structure has greater flexibility, i.e., the change of category in any "surface" in the classification system will not affect other "surfaces", and can add or delete "surfaces". Moreover, the classification structure of "surface" can be retrieved according to any combination of "surface", which is beneficial to computer information processing. Surface classification method has many advantages, such as extensive expansion, good structural flexibility, no need to determine the final grouping in advance, and it is suitable for computer management.

Disadvantage: The main disadvantage is that the configuration structure is too complex to be handled manually, and the encoding space cannot be fully utilized.

## 2.2 MasterFormat and UniFormat

In 1963, CSI of the United States and CSC of Canada jointly issued the CSI Architectural Code Format, which aims to be used for project archives management, cost management and compilation of organizational specifications. The earliest version of MasterFormat had 16 classifications. The first version was released in 1978. This version was mainly for housing projects and was maintained until 1995. Later, as many new products, materials and processes poured into the construction industry, machinery and pipelines became more and more complex, while more and more municipal and industrial projects are needed to be coded. None of this new information can be compiled into MasterFormat 1995. So in 2004, CSI released MasterFormat 2004, expanding the original 16 categories to 50 categories and grouping the 50 categories into six groups (Table 2-1).

MasterFormat Classification Form    Table 2-1

| MasterFormat Classification | |
|---|---|
| Procurement and Contracting Requirements Subgroup | |
| 00 Procurement and Contracting Requirements | |
| General Requirements Subgroup | |
| 01 General Requirements | |
| Facility Construction Subgroup | |
| 02 Existing Conditions | 03 Concrete |
| 04 Masonry | 05 Metals |
| 06 Wood, Plastics and Composites | 07 Thermal and Moisture Protection |
| 08 Openings | 09 Finishes |
| 10 Specialties | 11 Equipment |
| 12 Furnishings | 13 Special Construction |
| 14 Conveying Equipment | |
| Facility Services Subgroup | |
| 21 Fire Suppression | 22 Plumbing |
| 23 Heating, Ventilating, and Air Conditioning | 25 Integrated Automation |
| 26 Electrical | 27 Communications |
| 28 Electronic Safety and Security | |
| Sites and Infrastructure Subgroup | |
| 31 Earthwork | 32 Exterior Improvements |
| 33 Utilities | 34 Transportation |
| 35 Waterway and Marine Construction | |
| Process Facilities Subgroup | |
| 40 Process Interconnections | 41 Material Processing and Handling Equipment |
| 42 Process Heating, Cooling, Drying Equipment | 43 Process Gas and Liquid Handling, Purification, and Storage Equipment |
| 44 Pollution and Waste Control Equipment | 45 Industry-Specific Manufacturing Equipment |
| 46 Water and Wastewater Equipment | 48 Electrical Power Generation |

MasterFormat is a kind of building information classification method for materials and types of work. It can be used conveniently by development units in work breakdown, cost calculation and bidding. However, in the hands of investors and designers, the small categories of materials and processes will create obstacles in the aspects of scheme estimation, limitation design and dynamic cost control. Therefore, the American Institute of Architects (AIA) and the General Services Administration (GSA) developed two coding systems respectively according to the idea of classification of building elements. Later, they integrated their respective standards and jointly developed them in 1972, named UniFormat. The current UniFormat has two versions, one is UniFormat II released by ASTM, which was first released in 1993, the latest version is 2015; the other is UniFormat released by CSI and CSC (Table 2-2), the latest version is 2010. Both were developed from the original UniFormat released jointly by AIA and GSA. Both MasterFormat and UniFormat use hierarchical classification method.

Level 1 of UniFormat: Major Element Group    Table 2-2a

| | | |
|---|---|---|
| | A | Substructure |
| | B | Shell |
| | C | Interiors |
| | D | Services |
| | E | Equipment and Furnishings |
| | F | Special Construction and Demolition |
| | G | Site work |
| | Z | General ( CSI Particular) |

The First Three Levels of CSI UniFormat2010    Teble 2-2b

| Level 1 | Level 2 | Level 3 |
|---|---|---|
| A Substructure | A10 Foundations | A1010 Standard Foundations |
| | | A1020 Special Foundations |
| | A20 Subgrade Enclosures | A2010 Walls for Subgrade Enclosures |
| | A40 Slabs-on-Grade | A4010 Standard Slabs-on-Grade |
| | | A4020 Structural Slabs-on-Grade |
| | | A4030 Slab Trenches |
| | | A4040 Pits and Bases |
| | | A4090 Slab-On-Grade Supplementary Components |
| | A60 Water and Gas Mitigation | A6010 Building Sub-drainage |
| | | A6020 Off-Gassing Mitigation |
| | A90 Substructure Related Activities | A9010 Substructure Excavation |
| | | A9020 Construction Dewatering |
| | | A9030 Excavation Support |
| | | A9040 Soil Treatment |

## 2.3 Coding System for Classification of Information in the Construction Industry: Omniclass and Uniclass

### 2.3.1 OmniClass

In 1993, the technical report titled *Classification of Information in the Construction Industry* (ISO/TR 14177) pointed out that the classification scope of the original coding system could not cover all aspects of the construction industry, which was an incomplete system. This report proposes the framework of architectural information classification system based on surface classification, and defines some new architectural classification objects, such as facilities, spaces, design components, items, products, auxiliary tools, construction activities, etc., which lays the foundation for modern architectural coding system. In fact, two methods of building object classification are proposed: hierarchical classification and surface classification. In order to describe such complex objects as the construction industry, it is not enough to rely only on the hierarchical classification with the hierarchical ownership relationship, and there are more classification methods without the coherence, which need to be described through the surface classification, similar to the directed graph and undirected graph relationship in topology.

In 1996, based on ISO/TR 14177 and practical engineering experience, the ISO organization published an important standard: ISO 12006-2. A basic process model is proposed, which divides architecture into three categories: "construction process", "construction resources" and "construction achievements". This is actually an attempt to digitally describe the entire lifecycle of the construction industry in three dimensions (as shown in Figure 2-1).

Based on the ISO 12006-2 standard and referring to the inheritance resources related to the construction industry classification code, CSI led the establishment of OmniClass, the North American construction industry information model classification coding system. The code, drawn up by 17 organizations involved in the construction industry, began in 2000 and took six years to produce version 1.0.

OmniClass adopts the method of combining faceted system with hierarchical system, with a total of 15 classification tables. The 15 tables (shown in table 2-3) are classified by faceted taxonomy. It is only from different perspectives to understand the construction industry, these parts have a certain degree of relevance, but there is

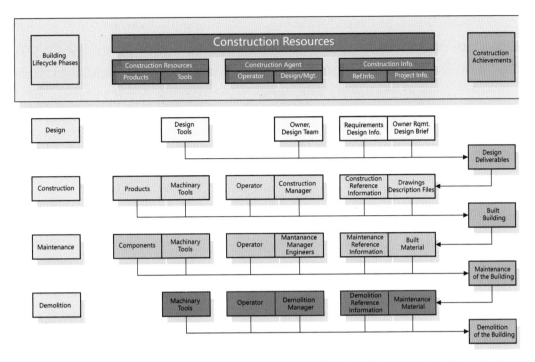

Figure 2-1 ISO12006-2 A digital description of the lifecycle of the construction industry

no subordinate relationship. Within each category, a hierarchical division is adopted, representing a classification method for building information.

On the one hand, this classification method can adapt to the complex and diverse characteristics of architectural information; on the other hand, it can fully inherit the results of existing traditional classification. Because the surface taxonomy is independent and not related to each other, it is saved as a separate file.

The inter-related OmniClass tables　　　　　　　　　　　　　Table 2-3

| Table 11-Construction Entities by Function | Table 32-Services |
|---|---|
| Table 12-Construction Entities by Form | Table 33-Disciplines |
| Table 13-Spaces by Function | Table 34-Organizational Roles |
| Table 14- Spaces by Form | Table 35-Tools |
| Table 21-Elements (includes Designed Elements) | Table 36-Information |
| Table 22-Work Results | Table 41-Materials |
| Table 23-Products | Table 49-Properties |
| Table 31-Phases | |

Although OmniClass is composed of 15 tables, the most commonly used table 23-product table is the classification code basis of components in Revit applications. The OmniClass Taxonomy file can be extracted from the Revit application installation directory (as shown in Figure 2-2). After opening it, the classification code of components can be seen starting from table 23 (as shown in Figure 2-3). However, only Table 23 can be embedded in Autodesk. The reason is that OmniClass classification and coding system is always based on the logical relationship between components, lacking of planning, planning, design, contract and implementation (completion) no matter it inherits form 21-elements of UniFormat, form 22-work results of MasterFormat or form 23-products of EPIC. The integration of the elements of "building logistics" in the operation and maintenance stages, and the information faults among the 15 tables produced by the facet method still exist. On the other hand, in different facet tables, codes are combined to describe complex building objects by logical operators. For example, descripting "the implementation phase of the children's library of slab-column structure column" needs the code combination of Table 31, Table 11, Table 23, to form "31-60 00 00+11-12 29 14+23-13 35 11 13 11". Such a lengthy 28-Bit code is only used to represent the "a specific object within the lifecycle of a construction project", which will lead to the decrease of computer storage efficiency and information transmission efficiency.

Figure 2-2  OmniClassTaxonomy    Figure 2-3  Component code in Revit

OmniClass is the combination of MasterFormat and UniFormat, and the results are shown in Figure 2-4.

| Phases | Preliminary Planning | Preliminary Design | Detailed Design | Tender | Procurement | Construction | Operation & Maintenance |
|---|---|---|---|---|---|---|---|
| Process | Conceptual cost plan | Detailed design & product selection | Detailed design & product selection | Price survey | | Procurement & change management | Property management |
| Coding Format | UniFormat | | | | | | UniFormat |
| | | MasterFormat | | | | | |

Figure 2-4  Basic combination of OmniClass

The Omniclass standard still lacks a clear idea of how the 15 tables fit together.

The BIM standards committee of the United States also pointed out in 2016 that due to the lack of a clear hierarchical framework structure of NBIMS, it could not form a unified business process of the lifecycle of the construction project. Therefore the standards were not implemented by software vendors, as shown in Figure 2-5.

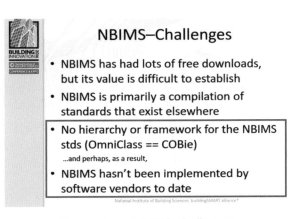

Figure 2-5  NBIMS challenges

### 2.3.2　UniClass

UniClass developed by CPIC in the UK is a unified classification code for buildings that combines faceted system and hierarchical system. The table composition of UniClass published in 2015 is shown in table 2-4 below. At present, NBS is responsible for updating and maintaining, and its classification and coding contents are integrated into the 15 tables of OmniClass, that is, UniClass and OmniClass are mutually inherited resources. The OCCS development committee can refer to and adapt the contents of UniClass according to actual needs, and this crossover also enables UniClass to realize further optimization for OmniClass reference.

UniClass is a dynamic classification system for the construction industry (The following table is updated to 28 October 2018)　　　Table 2-4

| Table | Status and revision information |
| --- | --- |
| Co-Complexes | V1.7, Published August 2018 |
| En-Entities | V1.10, Published October 2018 |
| Ac-Activities | V1.8, Published October 2018 |
| SL-Spaces/locations | V1.10, Published October 2018 |
| EF-Elements/functions | V1.3, Published August 2018 |
| Ss-Systems | V1.12, Published October 2018 |
| Pr-Products | V1.12, Published October 2018 |
| TE-Tools and Equipment | V1.5, Published August 2018 |
| PM-Project management | V1.2, Published August 2018 |
| Zz-CAD（CAD） | V1.0, Published July 2018 |
| FI-Form of information | Beta status |

No matter UniClass of CPIC or OmniClass of CSI/CSC, due to the imperfection of classification coding results, it can not uniformly serve all industry categories of construction engineering, nor can it meet the information integration requirements of the lifecycle of construction engineering. Compared with OmniClass, UniClass 2015, which is responsible for updating and maintaining by NBS, has many improvements. (1) the OmniClass table is published in PDF and Excel format, which is easier to retrieve and use to a certain extent compared with the paper code. (2) the 15 tables of OmniClass are not coordinated and unified, most of which are collected and sorted by all parties after independent development, while UniClass 2015 is a set of unified classification and coding system for the construction industry, and its terms, sorting and grouping structure among different tables are consistent. (3) coding structure, OmniClass table code consisting of no practical significance Numbers (e.g., 21, 22), the number code level from two levels to eight levels, while UniClass 2015 table code consisting of meaningful words abbreviations (En) such as activity (Ac), entity (En), and the code level ranges are mostly consistent at three to five levels (such as shown in Table 2-5, Table2-6).

UniClass Table Activities — Table 2-5

| Code | Group | Sub group | Section | Object | Title |
|---|---|---|---|---|---|
| Ac_05 | 05 | | | | Project management activities |
| Ac_05_00 | 05 | 00 | | | Strategy stage activities |
| Ac_05_00_10 | 05 | 00 | 10 | | Business case development |
| Ac_05_00_80 | 05 | 00 | 80 | | Strategic brief preparation |
| Ac_05_00_82 | 05 | 00 | 82 | | Strategic brief submission |
| Ac_05_10 | 05 | 10 | | | Brief stage activities |
| Ac_05_10_15 | 05 | 10 | 15 | | Cost estimate preparation |
| Ac_05_10_17 | 05 | 10 | 17 | | Cost estimate submission |
| Ac_05_10_29 | 05 | 10 | 29 | | Feasibility study preparation |
| Ac_05_10_31 | 05 | 10 | 31 | | Feasibility study submission |
| Ac_05_10_61 | 05 | 10 | 61 | | preliminary design preparation |
| Ac_05_10_63 | 05 | 10 | 63 | | preliminary design submission |
| Ac_05_10_65 | 05 | 10 | 65 | | Project brief and objectives preparation |
| Ac_05_10_67 | 05 | 10 | 67 | | Project brief and objectives submission |
| Ac_05_20 | 05 | 20 | | | Concept stage activities |
| Ac_05_20_15 | 05 | 20 | 15 | | Concept cost report preparation |
| Ac_05_20_17 | 05 | 20 | 17 | | Concept cost report submission |
| Ac_05_20_21 | 05 | 20 | 21 | | Concept design development |
| Ac_05_20_23 | 05 | 20 | 23 | | Concept design report preparation |
| Ac_05_20_25 | 05 | 20 | 25 | | Concept design report submission |
| Ac_05_30 | 05 | 30 | | | Definition stage activities |
| Ac_05_30_03 | 05 | 30 | 03 | | Agreement negotiating |
| Ac_05_30_10 | 05 | 30 | 10 | | Building regulation assessing |
| Ac_05_30_21 | 05 | 30 | 21 | | Definition design cost report preparation |
| Ac_05_30_23 | 05 | 30 | 23 | | Definition design cost report submission |
| Ac_05_30_25 | 05 | 30 | 25 | | Definition design development |
| Ac_05_30_27 | 05 | 30 | 27 | | Definition design report preparation |
| Ac_05_30_29 | 05 | 30 | 29 | | Drainage adoption agreeing |
| Ac_05_30_37 | 05 | 30 | 37 | | Highways adoption agreeing |
| Ac_05_30_60 | 05 | 30 | 60 | | Party wall notices agreeing |
| Ac_05_30_64 | 05 | 30 | 64 | | Planning preparation |
| Ac_05_30_85 | 05 | 30 | 85 | | Sustainability assessing |

continued Table

| Code | Group | Sub group | Section | Object | Title |
|---|---|---|---|---|---|
| Ac_05_40 | 05 | 40 | | | Design stage activities |
| Ac_05_40_85 | 05 | 40 | 85 | | Technical design cost report preparation |
| Ac_05_40_87 | 05 | 40 | 87 | | Technical design development |
| Ac_05_40_89 | 05 | 40 | 89 | | Technical design report preparation |
| Ac_05_50 | 05 | 50 | | | Build and commission stage activities |
| Ac_05_50_15 | 05 | 50 | 15 | | Contractor mobilizing |

UniClass Table Entities    Table 2-6

| Code | Group | Sub group | Section | Object | Title |
|---|---|---|---|---|---|
| En_20 | 20 | | | | Administrative, commercial and protective service |
| En_20_10 | 20 | 10 | | | Legislative entities |
| En_20_10_45 | 20 | 10 | 45 | | Governmental buildings |
| En_20_15 | 20 | 15 | | | Administrative office entities |
| En_20_15_10 | 20 | 15 | 10 | | Multiple occupation office buildings |
| En_20_15_70 | 20 | 15 | 70 | | Single occupation office buildings |
| En_20_20 | 20 | 20 | | | Secular representative entities |
| En_20_20_10 | 20 | 20 | 10 | | Buildings for representatives of nation states abroad |
| En_20_20_40 | 20 | 20 | 40 | | Local government buildings |
| En_20_20_50 | 20 | 20 | 50 | | National government buildings |
| En_20_20_70 | 20 | 20 | 70 | | Regional government buildings |
| En_20_45 | 20 | 45 | | | Motor vehicle maintenance and fuelling entities |
| En_20_45_50 | 20 | 45 | 50 | | Motor vehicle fuelling and charging entities |
| En_20_45_54 | 20 | 45 | 54 | | Motor vehicle servicing and repair entities |
| En_20_50 | 20 | 50 | | | Commercial entities |
| En_20_50_05 | 20 | 50 | 05 | | Auction buildings |
| En_20_50_22 | 20 | 50 | 22 | | Department stores |
| En_20_50_29 | 20 | 50 | 29 | | Financial and professional services buildings |
| En_20_50_50 | 20 | 50 | 50 | | Markets |
| En_20_50_53 | 20 | 50 | 53 | | Mixed use buildings |
| En_20_50_55 | 20 | 50 | 55 | | Motor vehicle sales entites |
| En_20_50_80 | 20 | 50 | 80 | | Shop units |
| En_20_50_85 | 20 | 50 | 85 | | Supermarkets |
| En_20_50_97 | 20 | 50 | 97 | | Wholesale buildings |
| En_20_55 | 20 | 55 | | | Postal communications entities |
| En_20_55_65 | 20 | 55 | 65 | | Post office buildings |

continued Table

| Code | Group | Sub group | Section | Object | Title |
|---|---|---|---|---|---|
| En_20_55_80 | 20 | 55 | 80 | | Sorting office buildings |
| En_20_60 | 20 | 60 | | | Military entities |
| En_20_60_02 | 20 | 60 | 02 | | Air force buildings |
| En_20_60_10 | 20 | 60 | 10 | | Army buildings |
| En_20_60_56 | 20 | 60 | 56 | | Navy buildings |
| En_20_65 | 20 | 65 | | | Law enforcement operational entities |
| En_20_65_09 | 20 | 65 | 09 | | Police buildings |
| En_20_70 | 20 | 70 | | | Judicial entities |
| En_20_70_40 | 20 | 70 | 40 | | Law court buildings |
| En_20_75 | 20 | 75 | | | Detention entities |
| En_20_75_10 | 20 | 75 | 10 | | Detention buildings |
| En_20_80 | 20 | 80 | | | Weapons training ranges |
| En_20_80_29 | 20 | 80 | 29 | | Firing range buildings |
| En_20_80_30 | 20 | 80 | 30 | | Exterior firing ranges |
| En_20_85 | 20 | 85 | | | Security entites |

In general, the project phase and project management table of UniClass 2015 reflect the lifecycle of the construction project, and all phases have the same status. In addition, UniClass is consistent with the system of ISO 12006-2:2015, which can be implemented into NRM1 (NRM——rules for expense estimation and itemized expense planning;NRM2——bill of quantities rules for engineering procurement; NRM3——mapping of operational and maintenance cost plans and procurement rules)and subsequent classification coding systems.

## 2.4 Problems of BIM Classification Coding

The first edition of MasterFormat in 1998 and the first edition of UniFormat II in 1993 are used for project archive management, cost management and preparation of organizational norms, program estimation, quota design, dynamic cost control. The first edition of American BIM standard was promulgated in 2007. Obviously, OmniClass was not born for BIM. Simply applying OmniClass to BIM is not successful.

Whether OmniClass in the United States or UniClass in the United Kingdom is used in BIM, it must be supported by the software provider vendors, and the reasons such as incomplete coding, imperfect coding or unusable coding may cause software vendors to abandon it and lose the significance of the coding.

Due to the complexity of architectural engineering, it is impossible to build a single application system that fully covers the lifecycle of a building. This application system should be composed of hundreds of different applications, and each engineering application is only based on a specific purpose to support specific phases of business work. In order to realize data interaction between different engineering applications based on BIM-centric data model, it is necessary to rely on unified data standards. In the construction Industry, the IFC (Industry Foundation Classes) architecture is currently the most comprehensive object-oriented data model, covering data definitions that meet all business requirements at all phases of the engineering design domain. The first version of the IFC standards was released in January 1997 by the IAI Industry Alliance for Interoperability (now buildingSMART International). In practical application, information sharing tools based on IFC need to be able to safely and reliably interact with data information, but the IFC standard does not define the different project phases, specific information needs between different project roles and applications, and the implementation of IFC-compatible application solutions encounters bottlenecks due to lack of specific information requirments definition. The application system cannot guarantee the integrity and coordination of interactive data. A BIM solution to this problem is to develop a set of standards that clearly define the actual workflow and the required interactive information, and this standard is the IDM standard (Information Delivery Manual). Obviously, the IDM is aimed at "completely covered the building the actual working process of the application system in lifecycle of a single application interaction information needed". Its goal is to standardize the information requirements for a specific stage of the lifecycle, and to provide the requirements to software vendors, mapping with the open data standards (IFC), and finally forming a solution. The first edition of American BIM standard and openBIM method released in 2007 take IFC+IFD+IDM as the basic standard to realize BIM.

In the actual implementation of BIM in this IFC standard intermediate data format, components are used as objects for modeling and information is transferred between different applications, forming the "component thinking" with IFC standard as the main interactive standard. This kind of "component thinking" has always dominated the BIM theory. However, up to now, for 20 years, IFC's "component thinking" theory has not been truly applied in practical engineering applications (Figure 2-6).

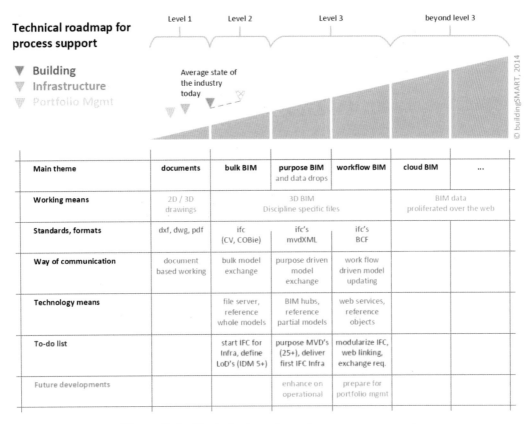

Figure 2-6  Technical roadmap for process support

A complete project has a large amount of information with high reuse frequency from planning to completion. For example, if there is no special requirement for a design, only use the design information to make sure that the concrete strength of the top structural column of a building is C30. In the subsequent construction drawings, the supplier of concrete and quality control personnel already have several sets of different evaluation criteria to fill in the concrete of the top structural column of the building with abundant information, such as cost, quality, pouring technology and maintenance time. There are many kinds of information, such as strength evaluation, inspection points, etc. The constructor, designer, constructor, supervisor, test inspector and supervisor will also describe this C30 from multiple perspectives. Faced with such a mature information network, the construction process of today's construction industry is still using traditional text to convey information, which requires us to rethink whether there is another better way to achieve the informatization of the construction industry.

# Chapter 3 The A&bCode System for Breakdown and Coding of Information in the Construction Industry

Pointed out by the technology report of *Classification of Information in the Construction Industry*(ISO/TR 14177), there are many working elements in the construction industry, and the relationship between these elements presents a complex state of intertwined.No matter what kind of organizational model is involved, the sum of each construction process, including planning phase, construction phase, operation and maintenance phase, and the demolition phase, is the same.Each process can be decomposed into activities that must be performed to drive the project forward.The action and business（A&b）is the major process in the classification analysis. The requirements for data are different for different professions and different tools at different phases when breaking down the construction industry information. In addition, the accuracy requirements of the results delivered at different phases are different; the results of different professional delivery are different; the results of different tools are delivered differently. Therefore, it is impossible to meet the data exchange needs of all relevant professional work simply based on a certain classification standard (for example, component-based or material-based).

In order to make different people work based on the lifecycle of the project, carry out various forms of data exchange, cooperate with each other, form synergy, and finally build the whole information system of the construction industry, a complete and reasonable coding system (A&bCode), which can unify the work results of different phases, different professions, different working methods, the exchange of file names and their exchange contents and format standards into an organic whole, is needed.

## 3.1 Breakdown and Classification

Segmentation: It is always split in both the parallel class and the hierarchy.

Breakdown:From top to bottom, it turned out to be a whole and each part of which can synergistically exert the overall function that each part (individual) does not have.

There are two important points for the breakdown, namely the integration and coordination between the constituent units and the constituent units.

Classification: From bottom to top, it turned out to be a whole.However, the elements that make up this whole arespatial, temporal or characteristic close to each other, and their overall characteristics are not necessarily greater than the sum of the characteristics of each element and play an integral role. The whole can be divided into elements, but the relationship is simple in space, time or characteristic.

There are also two indispensable key points for classification: one is classification, the other is abstraction.

## 3.1.1 Breakdown

Breakdown is the division of a whole object into individual components (individuals).

When conducting a single object analysis, the focus is not only to understand the whole, but also to understand the integrity of the body, the structure and composition of the internal, and the integration and synergy between the various components (individuals).Breakdown is the most important task in order to complete this work above.

The same as anatomy, breakdown is a process that goes from entirety to inside components through a variety of means, and then it further clarifies how these components work together. In other words, the various external manifestations of things ultimately involve how the internal units work together.

It is a representation that people sweat after exercise. However,we can understand how the various organs in the human body work together and ultimately lead to such a resultonly by going deep inside the objects.In this way, you can change from the appearance of things to the understanding of the internal mechanism of things.

Dismember an ox as skillfully as a butcher,anda master butcher sees through parts and joints of a ox without cutting.It is already engaged in the internal working mechanism of objects.

It should be noted that forming a complete breakdown structure or breakdown tree is just the first step of breakdown. More importantly, for each component or component (individual) that is decomposed, we also need to connect them according to the external representation. Only in this way can you find that the parts after the breakdown of objects are not isolated, but closely coordinated.

There are often two orders for Breakdown:

1) From static to dynamic: it can be decomposed firstly,and then study the synergy between each component (individual);

2) From dynamic to static: observe the dynamic operation process of objects firstly,find each specific unit component, and then consider how objects are scientifically decomposed.

These two methods of breakdown, we can easily think of direct anatomical breakdown if the research is as same as the human body structure research. However,you sometimes do not know which knowledge points or problem points should be decomposed for a new field of knowledge or problem.At this time, we must first study the dynamic formation process of the problem, and find the Breakdown unit.

No matter which breakdown order we choose, the two major problems must be solved: (1) decomposing into individuals and (2) coordination between individuals.

There is another important concept when objects are decomposed. For complex objects, whether it is actual objects or abstract knowledge, there may be a multidimensional structure. When there is such a multidimensional structure, we must learn to transform it into multiple two-dimensional structures, which are easier for us to perform visual analysis. This is a process of transforming things from complexity to simplicity.

It is much easier for us to understand the flattened structured graphics or image graphics in real life than the spatial model. For example, it is more difficult to learn stereo geometry than plane geometry. Therefore, we expect to convert the stereo geometry problem into a planar problem first. This is also the difficulty point. The university often offers mechanical drawing lessons for science and engineering. In those lessons, we can easily output projection views from three perspectives for a three-dimensional structure. Then here comes the second problem, if you are given by a plan with three perspectives, can you quickly conceive a specific three-dimensional image? It can be seen that the Breakdown is easier than the integration. Therefore a complete closed loop should be:

Stereo multidimensional → 2D or Single dimensional analysis → Integration analysis of dimensions → Aggregate reduction and closed loop.

This analytical thinking above can be used in many scenarios at work. Taking an application system as an example, when we analyze and design its architecture, we can see that there are multiple dimensions and perspectives in the architecture itself. Business processes, data, technology, physical deployment, and operational mechanisms may all be critical to the architecture perspective. If we want to make the framework clear, each of these individual dimensions must be analyzed and described one by one firstly. However, the core of forming a complete framework is not to describe each framework view separately, but how is it integrated between the various framework views, how individual single views can be integrated into a complete framework. The difficulty of the framework lies in the ability to integrate single views. Only by clarifying how to integrate can we give better analysis from static to dynamic, which is, from static models to internal dynamics.

The breakdown of any static object must consider the concept of hierarchy during decomposing, and should have the same granularity under different levels. There are two places that we most likely to make mistakes when decomposing, one of which is not in accordance with the Mutually Exclusive Collectively Exhaustive (MECE) rule. The MECE rule can solve a major problem without overlapping and not missing, and can effectively grasp the core of the problem and solve the problem. Another one is to decompose the hierarchical granularity problem. It will be a complete structural breakdown only if these two problems are solved. This kind of breakdown can also give better finalization based on the idea of the pyramid principle. It should be noted that the core reason for the ambiguity of the final presentation itself is the ambiguity of the structure at the time of breakdown. Moreover, the ambiguity of the deduction itself is also due to the errors in final induction .

The breakdown and presentation of a static object cannot really clarify the intrinsic mechanisms of objects. Therefore, it is more important to analyze the dynamics and operational mechanism of object itself. The dynamic analysis of objects is to regard things as a whole, and to study the evolution of the things themselves according to the timeline. After finishing this step, you will have a second question: what drives this dynamic change of objects? The dynamic development of an object is promoted by the external environment and objects on one hand, and by the interaction between the various components within the object on the other hand.

It is easy to decompose static state and dynamic state but difficult to combine them. That is, after static breakdown is completed, we need to consider the dynamic changes of things themselves and how static components interact and promote each other. From the dynamic view of the evolution of objects, it can be seen that the static components of which are active, and it is precisely because the dynamic development connects the static components.

An application system is broken down into multiple components when it is designed. But how exactly does a business process and function flow? The components need to interact with each other, which means that the dynamic business processes that are obtained need to be analyzed, how the components interact with each other, complete the closed loop and generate an output from the acquisition to a requirement. And all those questions must be figured out first. Static breakdown is often not an end, how statically decomposed components interact and integrate, complete and implement a dynamic goal, or ultimately complete aggregation into the original thing is the most critical.

Your research aim → Objects → Static breakdown → Dynamic analysis → Mixed dynamic and static analysis → Regression goal realization.

Do not forget your initial aim when you do any analysis and decision-making at any levels of thinking, or you will go further and further. Especially in the thinking, the more breakdown and analysis, the easier it is to continue to deepen and fall into the details. Do not forget your original intention, the goal-driven and closed-loop in thinking is something that must be realized at all in the thinking process.

### 3.1.2 Classification

Classification refers to the categorization of products according to type, grade or property.

Classification can sometimes be called categorization, which is more visible. Classification is a key method in product analysis. When facing tens of thousands of products, we cannot adopt different decision-making methods and action plans for different individuals.The most feasible method is to classify the products. After classification, we only need to adopt different decision-making methods and action plans for different objects.

When facing a product group, the first thing to consider is classification, that is, according to which key attributes of the product, the product group can be preliminarily sorted out.Take investment and financial management as an example,you may face many products and options.If you consider the risks and benefits, you can initially classify them into banking wealth management products, trusts, bonds, funds, P2P financing and other major categories.For fund classes, it may be divided into conservative bond investment funds and radical equity funds, and any fund product itself is a products unit of stock pools formed by multiple final stocks.

Classification is the first step, and after that you can choose two key dimensional propertiesfor matrix analysis according to the key characteristics and propertiesof the object itself. For example, risks and benefitscan be used as two key dimensions of the matrix forinvestment and financial management.

The purpose of abstraction is to study the common features of objects. We should first induce and refine abstract expressions from specific object, and then to deduct and expandnew objects in research.The abstract process is the process of induction, and the process of solving problems and making decisions is the process of deduction.

When we study a group of projects,it may include research and development projects, sales projects, company process optimization projects, operation and maintenance projects, etc., which is a basic classification. After this classification, we will start to study how each type of project should be managed and what are the key characteristic attributes of such projects.After the study, we will find that regardless of the type of project, the key attributes of project management such as planning, task tracking, milestones, and reporting mechanism are common. Then it also needs to be further abstracted into a company-level basic project management methodology, defining the most basic standard specification system, and ultimately enabling a complete view of project management for all of the company's projects. All kinds of projects have common contents and special contents, common basic contents must be consistent, and special contents can be customized according to project characteristics.

Further abstraction is needed to study a complete product. Abstraction can also be understood as induction, which is to find common features and attributes of different products, from the final product and product to the abstract thinking expression.

For the classification of products, we need to find the common properties of objects, and then classify them according to the different contents of these common properties. These properties may be static properties, such as whether the product is circular or square. They also may be dynamic properties, such as linear motion or curved motion. All of which can be used as attributes to classify products.

The analysis above shows that the process of breakdown is a top-bottom process,and the process of classification and abstraction happens to be a bottom-up process.

The purpose of breakdown is to decompose from the whole to the individual,and through the analysis of the individual to understand the inner working mechanism of the object.

The purpose of classification is to achieve unified decision-making and action on abstract categories through classification and abstraction from the individual to the whole.

Therefore, the classification of the AEC industry information without breakdown of the AEC industry information is bound to be a huge and uncontrollable classification .

## 3.2 Work Breakdown Structure (WBS)

The first step in achieving the digital expression is to figure out what the final delivery of the digital expression is. Trying to map this final deliverable requires the work breakdown structure.

### 3.2.1 WBS

WBS is a scope management tool originally proposed by the U.S. Department of Defense. A Guide to the *Project Management Body of Knowledge* (6th edition) defines the work breakdown structure as "work breakdown structure is the hierarchical breakdown of all work areas that a project needs to implement in order to achieve the project objectives and create the required deliverables. Each level down in the work breakdown structure represents a more detailed definition of the project work." In project management practice, the scope of the project is defined by the WBS, and the final deliverables are mapped out upon the completion of WBS. The WBS is always at the center of the planning process and is an significant basis for developing schedule plans, resource requirements, cost budgets, risk management plans, and procurement plans. At the same time, the WBS is an important foundation for controlling project changes.

Creating a work breakdown structure (Figure 3-1) is the process of breaking project deliverables and project work into smaller, more manageable components.

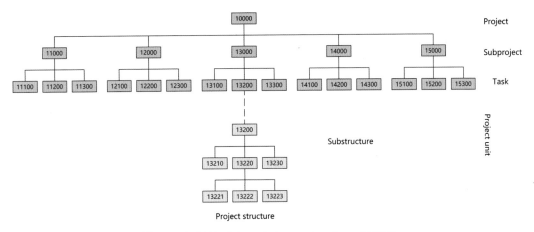

Figure 3-1 Work breakdown structure (WBS)

Work——tasks that produce tangible results; Breakdown——a hierarchical structure that is subdivided and categorized progressively; Structure——organizing the parts according to a certain pattern. According to these concepts, WBS has corresponding constituent factors corresponding to it:

(1) Structured code

Code is the most significant and critical component of WBS. First, code is used to thoroughly structure WBS. Through the coding scheme, we can easily identify the hierarchy, grouping categories and features of WBS elements. Due to the development of modern computer technology, code actually makes WBS information closely related to organizational structure information, cost data, schedule data, contract information, product data, and report information.

(2) Work package

The work package belongs to the lowest level element of a WBS, and the general work package is the smallest "deliverables" that can easily identify the information of activities, costs, organization, and resource to accomplish it. For example, the pipeline installation work package may contain several activities such as the fabrication and installation of pipeline supports, pipeline connection and installation, and strict inspection; including transportation/welding/pipeline manufacturing labor costs, pipeline/metal accessories material costs and other costs; reports/inspection results and other documents generated during the process; as well as the assignment of work teams and other responsibility package information and etc. Based on the above viewpoint, a WBS for project management must be decomposed to the work package level to enable it to be an effective management tool.

(3) WBS elements

WBS elements are actually the "nodes" of the WBS structure, the popular understanding is the "square boxes" in the "organization chart" that represent the "deliverables" that are independent and have affiliation/aggregation relationships. After decades of summing up, most organizations tend to believe that the WBS structure must be relevant to the project goals and must be oriented toward the end product or deliverable, so the WBS elements are better suited to the noun composition that describes the output product. The reason is obvious. Different organizations and cultures use different methods, procedures and resources to accomplish the same work, but their results must be the same and they must meet the requirements. Only by capturing the core deliverables can the project be controlled and managed effectively; on the other hand, only by identifying deliverables can the methods, programs, and resources used by internal/external organizations to accomplish this work be identified. The work package is the lowest level WBS elements.

(4) WBS dictionary

Normalization and standardization of management have always been the goal of pursued by many companies, WBS dictionary is such a tool. It is used to describe and define the documents that work in the WBS elements. The dictionary is equivalent to the specification of a certain WBS element, that is, the work that the WBS element must complete and the detailed description of the work; the description of the work accomplishments and the corresponding normative standards; the superior-subordinate relationship of elements and the input-output relationship of elements accomplishments. At the same time, the WBS dictionary has a huge normative role in clearly defining the scope of the project, which makes the WBS easy to understand and acceptable to participants (such as contractors) outside the organization. In the AEC industry, the bill of quantities specification is a typical work pack-level WBS dictionary.

The creation of WBS should be carried out in accordance with the actual work experience and the method of system work, the characteristics of the project and the requirements of project managers. Its basic principles are as follows:

1) WBS should follow the MECE (Mutually Exclusive Collectively Exhaustive) principle, that is, when decomposing a task, it should be independent of each other and completely exhausted;

2) WBS should follow the SMART principle, that is, the breakdown of a task must have five conditions: specific, measurable, attainable, relevant, and time-bound. There must be departments and people responsible for each task, and there must be main chief, specific to individuals, rather than assigned to groups of several people;

3) Each phase of the project should be able to distinguish between different responsibilities and different work content, and should have a high integrity and independence;

4) The principle of visualization, each refinement task can be seen hierarchically; components can be moved to different positions to facilitate the layout of WBS;

5) It can meet the requirements of project target management, and can conveniently apply the schedule, quality, cost, contract, information and other means;

6) WBS should be decomposed into 4~6 layers, and the unit cost of the lowest level of work package should not be too large and schedule should not be too long.

### 3.2.2 SWBS and PSWBS

(1) SWBS

Summary Work Breakdown Structure (SWBS) is the guidance and strategic work breakdown structure. The breakdown structure has only following three levels:

Level 1: SWBS system is used in different phases of each sub-industry of the AEC industry;

Level 2: the subsystem at different phases;

Level 3: the sub-system subordinates to subsystem, that is, the basic unit of the AEC industry.

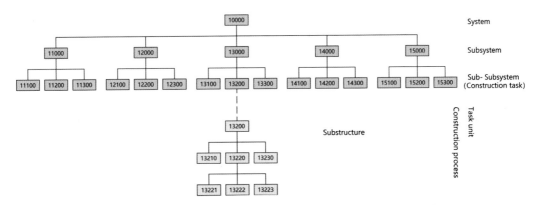

Figure 3-2 Summary Work breakdown structure (SWBS)

(2) PSWBS

Project summary work breakdown structure (PSWBS) is a work breakdown structure tailoring SWBS for a specific project. For specific projects, the first three levels of PSWBS in Figure 3-1 can be tailored from Figure 3-2.

### 3.2.3 Systematic Classification

Thinking reflects the object of knowledge by means of concepts and categories. These concepts and categories exist in the human brain in a certain form of framework, namely, the thinking structure.Different categories and concepts can be organized together and form a relatively complete idea through these frameworks, and then they will be able to understand, master and achieve the purpose of understanding by us. Therefore, the thinking structureis not only a cognitive structure of human beings, but also the ability structure of people to use the categories and concepts to grasp the object.

Logical thinking refers to the way or form of linking and organizing the contents of thinking. Logical thinking, as is a high-level form of thinking, refers to the way of thinking that conforms to the relationship between objects in the world (in accordance with the laws of nature). What we call logical thinking mainly refers to the way of thinking that follows the traditional formal logic rules. We usually consider "logical thinking" as "Abstract thinking" or "Closed eye thinking". Logical thinking is a kind of thinking that is certainty, consistent, organized and valid.Thethinking modes such as concept, judgment, and reasoning, and the thinking methods such as comparison, analysis, synthesis, abstraction, and generalization are used in logical thinking.The degree to which these forms and methods of thinking are mastered and applied is also the ability of logical thinking.

SWBS is a logical thinking expression.

Dialectical thinking is a reflection and conforms to the dialectical development process of object and its regularity. It is also a certain degree of understanding and application of objective dialectics and dialectic of cognitive processes.The characteristic of dialectical thinking is to examine from the internal contradictory movements of the object and from the interrelationship of its various aspects in order to understand the object in its entirety and in essence.Dialectical thinking uses logical categories and systems to grasp specific truths.Dialectical thinking is different from the isolated metaphysical thinking and the formal logical thinking, and it is the object of dialectical logic research.The historical development of human dialectical thinking has gone through a process from spontaneous to conscious.

Dialectical thinking refers to the way of thinking that understands things from the perspective of changing development. It is usually considered as a way of thinking that is contrary to logical thinking.In logical thinking, objects are generally "either this or the other" and "either true or false" . However, in dialectical thinking,objects can be "also this and the other" and "also true and false" at the same time without affecting the normal conduct of thinking activities.Dialectical thinking is a kind of worldview. Everything in the world is interconnected and affects each other.Dialectical thinking is a kind of thinking that further understands and perceives the world based on the objective connection between all things in the world, and feels the relationship between human being and nature in the process of thinking, and finally gets some conclusions. It is required to see the problem in a dynamic development perspective when you are observing and analyzing in dialectical thinking.

The task unit or substructure in Figure 3-1 and Figure 3-2 contains the specific activity process of completing the task (construction), belonging to the category of dialectical thinking.

System classificationmethod is an information breakdown method that uses the "limited hierarchical classification method" and the individual object structure coding, and express the whole object by different individuals with high information integrity. When we apply the System classification method to WBS, and the "limited hierarchical classification method" will control the hierarchical classificationwithin three layers, which we called SWBS. The SWBS not only retains the information capacity, the clear level, the stronglogic, conforms to the habit of traditional application, but also suitable for both manual operation and computer management.It takes advantage while avoiding all the shortcomings of thehierarchical classification method.

According to thesystem classification method, Figure 3-1 is divided into two parts, which arethe task and the project task unit above and below.In the AEC industry, "task" means that the professional work and construction methodsare basically fixed, and this part can be decomposed by "logical thinking" , which is certain,consistent,organized and well-founded. We will make the task and its upper part conform to the natural law through the thinking modes such as concept, judgment, and reasoning, and the thinking methods such as comparison, analysis, synthesis, abstraction, and generalization.

The lower part of Figure 3-1 is the "Task Unit" section. The "Task Unit" is the personnel, machinery, materials, methods, and environment used to complete the task. It is a variable, which means the construction process of the "task" should be analyzed by "dialectical thinking". We should investigate the object from the internal contradictory movements and the interrelationship of its various aspects, in order to understand the object in its entirety and in essence. All the changing rules of these interrelated and mutually influential objects will be unified in the "task".

## 3.3 Pattern

### 3.3.1 Pattern Thinking

Pattern is not well defined, but one of the most well-known definitions of the term was coined by architect Christopher Alexander, that is, "each pattern describes a problem that occurs over and over again in our environment, and then describes the core of the solution to that problem, in such a way that you can use this solution a million times over, without ever doing it the same way twice."

According to the definition above, the pattern consists of two basic elements: the problem and the core solution to that problem. Looking at the problem first, the purpose of the pattern is to decrease the cost of fixing problem by adopting reusable solutions, if that problem can occur repeatedly. On the contrary, a problem and a solution cannot be a pattern if the occurrence of that problem is occasional. As for the core solution of the problems, the word "core" denotes that the core of the solution is identical with some minor differences, even though the same problem occurs in different contexts. However, if the core of the solution to the problem is the difficult issues of solving problem, the influence of various background can be settled easily, so it is still compliant with the objective of the pattern.

The critical elements of the pattern are that they all come from practice. It is necessary to observe the working process of different people and find out the logic of scheme design so that "these core solutions to the problem" can be figured out.

Each pattern is relatively independent, interact with but not isolated from each other. The main value of the pattern is not how much new content or form it can provide, but it can be surely telling us which solutions are effective and with pattern thinking in the process of practice (we will also emphasize that the pattern comes from practice), and assist in reusing knowledge.

### 3.3.2  Application of Pattern Thinking in the Construction Industry

We have classical and local methods of component breakdown, calculation, implementation and resource combination in the field of construction. These classical reusable solutions always reflect practice-based pattern thinking. The concepts above all have relatively stable form and implementation logic, which are the fundament of breakdown and coding of the construction Industry information.

IDM is the interactive information (MVD) required for a single task (application) of a practice workflow in an application system that completely covers the entire lifecycle of building. As a result, IDM is a systems engineering. Albert Einstein once said: "We cannot solve our problems with the same thinking we used when we created them, but look at them from a higher level". Different sub-industries of the construction industry in China are usually with their own fixed workflow, and different workflow nodes need task applications. The workflow satisfies the definition of "pattern".

"Pattern Thinking" is to program a complete "distributed BIM database and its IDM/MVD system" in combination with the workflow of the AEC Industry, to substitute distributed database for the single format of "Component Thinking" ("centralized") BIM database, to replace BIM modeling application with distributed software system, and to replace single IFC standard exchange with end-to-end various standard exchange requirements. A new BIM" Pattern Thinking" is forming, which is a dynamic BIM system composed of several different but interrelated functional components. It is a kind of BIM implementation approach that break down the whole into parts, break the whole down individually, and realize simplification, improvement, multidimensional participation, and diversification of BIM implementation. It implies the mystery of the "dividing and ruling" of the art of war. The P-BIM theory proposed by *National Unified standard for building information modeling* is essential a Pattern Thinking.

Compared with the rapid development of the Internet, the "Component Thinking" of BIM develops slowly and conservatively. The space capacity provided by existing Internet services is not omnipresent, which means the development space left for component thinking is still huge. However, another opportunity of informatization of the construction Industry may be missed again if we continue follow the traditional ways in the idea and technology. In the past 20 years, BIM has focused on data modeling of "Component Thinking ", which is only a tool of visualization and independent of spatial computing. In the era of big data nowadays, the use of data to support decision-making is the essential value of BIM. The core of BIM aided decision-making is synthesis and efficiency, and the component thinking becomes an unsolvable problem due to its inherent defects. In contrast, HIM matrix grid based on pattern thinking is the technical roadmap of Internet service. Grid is the basic unit of spatial computing, information carrying and computing. Tile Map, space search, real-time traffic and car booking are all grid without exception. Because the grid is recorded in the computer as the unified code, calling the code for various operation is a universal method of IT, and varied data optimization approaches can be adopted to resolve massive data and access of BIM system. Pattern Thinking contains Component Thinking.

### 3.3.3 Two significances of Pattern Application

1) Pattern refers to the core knowledge system abstracted and sublimated from production and life experience. Pattern is actually methodology for addressing a certain type of problem. To summarize the approaches for solving certain problems to the theoretical level, that is the pattern. Pattern is a kind of favorable guidance under which you can finish the tasks with a reasonable design scheme and get more from less. And you will find the best way to resolve the problem.

2) Pattern is also a definite approach of thinking in the sense of epistemology. It is the abstraction and sublimation of people's experience accumulated in practice. Simply speaking, it is some laws discovered and extracted from recurring events, which is a refined summary of the experience of settling problems. In other words, there may be a pattern as long as it is a recurring event.

### 3.3.4 "System" Pattern Thinking in the AEC Industry

*The Art of Thinking in Systems* of Steven Schuster provided the definition of System, that is, "System is a whole composed of many parts, each part is correlated with each other and have a common objective as a whole. Human bodies, schools, companies and countries are all systems."

Breakdown of the construction industry information is a classical pattern system of long-term practice of human beings. Conventional practice-based patterns are discrete, and it is hard to be integrated logically or used in a breakthrough way due to its insufficiency of tier architecture. However, the breakdown of the construction industry information presents the implementation boundary of these patterns, defines the relationship between them, and takes into account the objective between individuals and systems. Consequently, the breakdown of the construction industry information is an application of the system theory.

## 3.4  WBS in the AEC Industry

The AEC industry is a huge system, and the AEC industry informationization is to serve the AEC industry. Only by thoroughly analyzing the status and mode of engineering technology and management in the AEC industry can we combine informationization with engineering practice and make useful information systems. In engineering practice, we divide the AEC industry into sub-sectors such as building engineering, road engineering, public civil engineering, railway engineering, and etc. Different sub-sectors are independent and interrelated.Each sub-industry project has its own unique SWBS, which is suitable for PSWBS of any project. Work breakdown structure in the AEC industry can be seen in Figure 3-3.

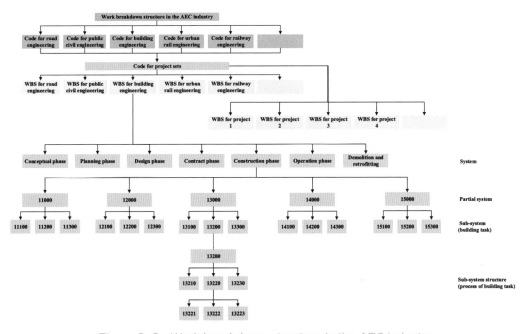

Figure 3-3  Work breakdown structure in the AEC industry

In Figure 3-3, the AEC industry was first decomposed into different sub-sectors. Project sets of different sub-industries (collection of multiple projects) are named. Subsequently, project sets are decomposed according to the commonality. The specific PSWBS of the project can be attained through the cutting of SWBS.

Third layer subsystem (construction tasks), the second layer of building, the first layer of building (Figure 3-2) and the organizational breakdown structure (OBS) corresponding to each task can be expressed in a column matrix:

$$\left\{ \begin{array}{c} \text{SWBS} \\ \text{The third layer} \\ \text{Building code} \\ 11100 \\ 11200 \\ 11300 \\ 12100 \\ 12200 \\ 12300 \\ 13100 \\ 13200 \\ 13300 \\ 14100 \\ 14200 \\ 14300 \\ 15100 \\ 15200 \\ 15300 \\ \ldots\ldots \\ \text{The second layer} \\ \text{Building code} \\ 11000 \\ 12000 \\ 13000 \\ 14000 \\ 15000 \\ \ldots\ldots \\ \text{The first} \\ \text{Building code} \\ 10000 \\ \ldots\ldots \\ \text{Corresponding OBS code} \\ \text{for each layer} \\ 90000 \\ 91000 \\ 92000 \\ \ldots\ldots \end{array} \right\} \quad (3\text{-}1)$$

## 3.5 MBS

The definition of model breakdown structure (MBS): Grouping project building information model elements with independent deliverables information. It summarizes and defines each level of work on the project building information model from the master-slave database structure system to the distributed database structure.

Summary model breakdown structure (SMBS) corresponds to SWBS.

Digital Twin is a digital representation of physical products. We can see what can happen to an actual physical product based on this digital product. Related technologies include augmented reality and virtual reality. In the process of design and production, Digital Thread can simulate the parameters of the analysis model, pass it to the full three-dimensional geometric model defined by products, and then transfer it to the digital production line to process into a real physical product. Through online digital detection, the measurement system is reflected in the product definition model and then fed back into the simulation analysis model.

From the perspective of dialectical thinking, since the creation of WBS is a process of decomposing project deliverables and project work into smaller, more manageable components; then the creation of MBS is to decompose the overall building information model according to WBS, and decompose project information into smaller, more manageable components of information. The digital representation of WBS can be realized by corresponding to the task layer of the work breakdown structure.

Mapping: Assuming that $A$ and $B$ are two sets, if according to some correspondence rule $f$, for any element $x$ in set $A$, there is a unique corresponding element $y$ in set $B$, then such correspondence is called the mapping from set $A$ to set $B$. record as $f: A \rightarrow B$ ($y$ is the image of $x$, and $x$ is the original image of $y$).

The US National Building Information Model Standard Project Committee has the definition of BIM: "Building Information Modeling (BIM) is a digital representation of physical and functional characteristics of a facility". The mapping from the third layer "building" in Equation (3-1) to MBS distributed database is the most detailed information definition of BIM: "A BIM is a shared knowledge resource for information about a facility forming a reliable basis for decisions during its lifecycle; defined as existing from earliest conception to demolition".

WBS summarizes and defines tasks in each hierarchical breakdown of the total scope of work. MBS summarizes and defines task information in each hierarchical breakdown of the total scope of work. Since MBS is a distributed database, the cooperative database is located above the task in order to ensure the coordination and consistency of the third tier distributed database, namely the mapping from the second layer "building" in Equation(3-1)to the collaborative system database.Similarly, the first level "system building" completely maps to the system database.

The NBIMS defines that "a basic premise of BIM is the collaboration of different stakeholders at different phases of the lifecycle, including inserting, acquiring, updating and modifying information in BIM to support and respond to the responsibilities of stakeholder". OBS of a project is about the internal organization of the project, not the relationship between organizational elements and its parent organization, matrix or other organization. It is an internal organization chart of a project constructed in a similar way to the work breakdown structure. OBS describes a specific organizational unit responsible for each project activity. It is a project organization diagram that links work packages with related departments or units hierarchically and systematically. OBS is an evolutionary approach. It is a way to break down people at all levels within an organization. OBS is not the same as internal organizational breakdown system. For example, although a person is at a lower level of the organizational system, he or she may need to understand the overall situation, so he or she may need to be at a higher level of OBS. In addition, OBS includes organizations of project participants and can even be extended to "project stakeholders". Project stakeholders (such as owner, government, design institute, general contractor and subcontractor) use their respective project management platforms to obtain and create BIM data, the corresponding database of OBS organization and management is generated, and the corresponding mapping between OBS and the management database of project stakeholders is made at each level.

Therefore, the mapping between SWBS in equation(3-1) and SMBS is:

## 3.6 Distributed Functional Modeling Application (P-BIM Functional Application)

### 3.6.1 Nonlinear Characteristics of the AEC Industry

The problem with definite causation is a linear problem——since there is a result,

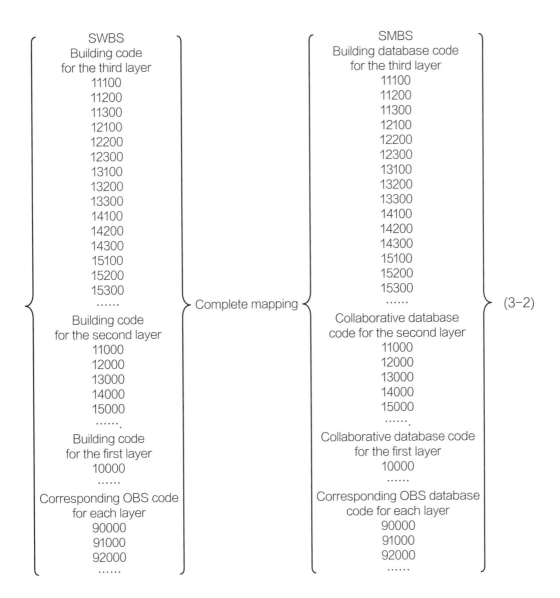

there must be a reason. As long as the reason is solved, the problem can be solved. For instance, the cell phone is out of power, we just need to charge it; the tires of automobiles are aging, we just need to replace a new qualified tire. The linear problem is simple and clear, and the causal relationship is clear.

The problems with complex or ambiguous causality are all non-linear problems. The problems of construction projects are mostly non-linear. For example, the concrete quality of a pier column is not good. The reason may be that the concrete of the pier column is not qualified. The reason may be that the operation violates the rules when pouring (adding water or waiting too long). The reason may also be that the maintenance is not in place, or even the extreme abnormal weather.

The AEC industry system is essentially an uncertainty and non-linear system which does not satisfy the superposition principle.

### 3.6.2 Traditional Modeling Method for Nonlinear Systems

Modeling and control of complex systems have always been a hot issue in the control field. Traditional global modeling methods generally have problems such as too many undetermined parameters and difficult to determine the model structure and difficulty in determining the model structure. If the object of concern is a high-dimensional space problem, there is also the problem of vulnerability to "dimension disaster". At the same time, the effect of most methods is related to the user's experience. In practice, in order to obtain high-precision estimation, it usually takes a lot of time to choose the appropriate modeling method, determines the appropriate model structure or model set, and adjust the algorithm and parameters repeatedly. However, even if high-precision models are obtained, they are often too complex to be further analyzed and applied.

In the field of control engineering, such a kind of complex non-linear system exists widely: due to the effect of non-linear factors, the operating mode of the system is too complex, or the working environment is very bad, resulting in a strong uncertainty of the structure and parameters of the system with the change of working conditions. At this time, it is difficult to establish a single high-precision mathematical model by using the traditional global modeling method. Because the effect of most modeling methods is closely related to the user's experience, model identification is complex, computational complexity is large, time-consuming is long, and it is difficult to apply in practice. Sometimes, even if the model of the system is obtained, it is difficult to carry out stability analysis and controller design because the model itself is too complex. In order to solve this problem, experts and scholars in the field of control have made a lot of theoretical and technical explorations, and finally found that the multi-model method based on the idea of "divide and rule" is an effective way to solve the problem.

### 3.6.3 Multi-model Modeling Theory

Since the emergence of the multi-model method in the 1970s, it has gone through more than 40 years of development, and has achieved extensive success in theory, technology and application. This method takes the principle of breakdown-synthesis as the natural way to solve the problem, decomposes the modeling and control problems of complex systems according to certain deterministic criteria, and then establishes local models and controllers respectively. Because local models are often linear, mature linear system theory can be used to design controllers. After the problem is decomposed and the solution is obtained, the local models or controllers can be coordinated according to the corresponding synthesis criteria or scheduling mechanism, and the control of the original complex nonlinear system can be realized while ensuring the stability. This modeling and control method can transform complex non-linear problems into problems that can be solved by mature and relatively simple theories. It is of great significance to simplify controller design and improve the control performance of systems in complex environments. At present, multi-model methods mainly include multi-model modeling method, multi-model control method and interactive multi-model filtering method. However, no matter what kind of methods it is, there are two key problems to be solved: one is to select the appropriate model set under the specific breakdown criteria; the other is to select the appropriate synthesis criteria to ensure the stability of model scheduling.

The most important methodology of multi-model modeling theory is "breakdown-synthesis". Through the application of "breakdown-synthesis" modeling method, the whole non-linear system is decomposed, and several simple local models are used to approximate the original system. This modeling framework has two advantages:

1) It has the ability to analyze the characteristics of each local model using mature linear system modeling strategies.

2) For complex nonlinear systems, complex model structures must be chosen to approximate the whole system. As long as appropriate learning strategies and compensation mechanisms are adopted, both reasonable approximation accuracy and computational complexity can be achieved.

Therefore, the idea of using multiple local models to approximate the dynamic characteristics of the system and designing adaptive controllers based on each local model is proposed. Multi-model method has the characteristics of intelligent control. It combines classical modeling and control methods with advanced control ideas. Its basic principle is simple, its algorithm is simple and easy to implement.

The piecewise affine/linear multi-model method divides the state space into a finite area of surface angle, and describes the dynamic characteristics of the system with a piecewise affine/linear function in each area of surface angle. Based on the idea of multi-model breakdown-synthesis, piecewise affine/linear multi-model divides the space of the system into finite subintervals, which are represented by a piecewise affine/linear local model, and then describes the dynamics of the system by switching.In addition to approaching any sufficiently smooth nonlinear function with any accuracy, this kind of model can also be used to approximate non-linear systems with discontinuous characteristics. In addition, it is closely related to hybrid systems and proved to be equivalent to hybrid systems of hybrid logic dynamic systems, linear complementary systems, extended linear complementary systems and maximum and minimum positive coordinate systems. Therefore, piecewise affine/linear multi-model can be used to identify hybrid systems.

The most direct way to solve the piecewise affine/linear system identification problem is to select the cost function according to the model structure and parameters, and then solve it directly by the numerical method or optimization algorithm.

With the increasing complexity of the controlled objects, more and more modern systems have the characteristics of strong nonlinearity, strong coupling and wide operating range. Using a single model to describe the global system is often too complex to meet the actual needs of the controller design because of the complexity of the modeling results. In addition, even if a satisfactory global model can be obtained. The identification process may also be too complicated and too many data rules are needed, which results in too much computation and time-consuming to be applied in reality. Therefore, it is necessary to fully mine the useful information hidden in the sample data of the control object and find other effective modeling methods.

The un-success of "combining and governing" has prompted people to turn to "dividing and governing". Many local models of controlled objects are used instead of a single global model to model separately. Finally, the local multi-models are synthesized into a whole by some means to approach the global model. This is the principle of "breakdown-synthesis" in the modeling of non-linear systems. The multi-model modeling method based on the breakdown-synthesis principle can be briefly described as three steps:

Firstly, according to some breakdown criteria, the working area of the whole system is divided into several intervals, and the whole interval and the characteristic variables used to represent the interval are defined.

Secondly, the local model structure on each interval is selected and its parameters are identified.

Finally, according to some scheduling mechanism or performance index, the local models are combined to get the solution of the original problem.

SMBS is based on the idea of multi-model modeling. The application is a tool for assisting tasks. For WBS, there must be an application for each task (corresponding application function in WBS system); for SMBS distributed database, there must be a modeling application for each database. Therefore, the coding of distributed function modeling application (P-BIM application function) completely maps to WBS code and MBS code (Equation 3-3).

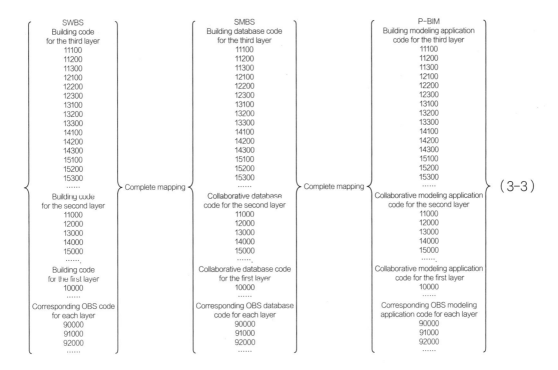

## 3.7 P-BIM Functional Application Information Exchange Standard

The basic requirement of American BIM standards for BIM is that "BIM is a shared digital representation based on collaborative performance disclosure standards". Every P-BIM functional application must be equipped with a *XXX P-BIM Application Function* and Information Exchange Standard. For example, the Standards for Application Functions and Information Exchange in Architectural Design P-BIM corresponds to the MVD of all relevant architectural design applications delivered by openBIM to other software, as shown in Figure 3-4。

Figure 3-4　P-BIM Exchange Standards vs MVD

Therefore, the standard code of functional modeling applications exchange corresponds to the function of P-BIM applications, forming the complete mapping relation of SWBS/SMBS/P-BIM application function/P-BIM application exchange standard coding (Equation 3-4)。

## 3.8　Construction Information Breakdown Coding System

Equation (3-4) presents the encoding based on SWBS, corresponding to SMBS, P-BIM applications, and P-BIM application exchange standard coding. The construction industry informatization provides services the basic work unit of the construction industry. As shown in Figure 1-4, the objective of information exchange is Exchange supporting an action or business. Actually, each action in P-BIM application system includes business, such as construction of cast-in-place piles. Therefore action and

| SWBS | | SMBS | | P-BIM | | P-BIM |
|---|---|---|---|---|---|---|
| Building code for the third layer | | Building database code for the third layer | | Building modeling application code for the third layer | | Building application information exchange standard code for the third layer |
| 11100 | | 11100 | | 11100 | | 11100 |
| 11200 | | 11200 | | 11200 | | 11200 |
| 11300 | | 11300 | | 11300 | | 11300 |
| 12100 | | 12100 | | 12100 | | 12100 |
| 12200 | | 12200 | | 12200 | | 12200 |
| 12300 | | 12300 | | 12300 | | 12300 |
| 13100 | | 13100 | | 13100 | | 13100 |
| 13200 | | 13200 | | 13200 | | 13200 |
| 13300 | | 13300 | | 13300 | | 13300 |
| 14100 | | 14100 | | 14100 | | 14100 |
| 14200 | | 14200 | | 14200 | | 14200 |
| 14300 | | 14300 | | 14300 | | 14300 |
| 15100 | | 15100 | | 15100 | | 15100 |
| 15200 | | 15200 | | 15200 | | 15200 |
| 15300 | | 15300 | | 15300 | | 15300 |
| …… | Complete mapping | …… | Complete mapping | …… | Complete mapping | …… |
| Building code for the second layer | | Collaborative database code for the second layer | | Collaborative modeling application code for the second layer | | Collaborative application information exchange standard code for the second layer |
| 11000 | | 11000 | | 11000 | | 11000 |
| 12000 | | 12000 | | 12000 | | 12000 |
| 13000 | | 13000 | | 13000 | | 13000 |
| 14000 | | 14000 | | 14000 | | 14000 |
| 15000 | | 15000 | | 15000 | | 15000 |
| …… | | …… | | …… | | …… |
| Building code for the first layer | | Collaborative database code for the first layer | | Collaborative modeling application code for the first layer | | Collaborative application information exchange standard code for the first layer |
| 10000 | | 10000 | | 10000 | | 10000 |
| Corresponding OBS code for each layer | | Corresponding OBS database code for each layer | | Corresponding OBS modeling application code for each layer | | Corresponding OBS application information exchange standard code for each layer |
| 90000 | | 90000 | | 90000 | | 90000 |
| 91000 | | 91000 | | 91000 | | 91000 |
| 92000 | | 92000 | | 92000 | | 92000 |
| …… | | …… | | …… | | …… |

(3-4)

business（A&b）expression is more appropriate.

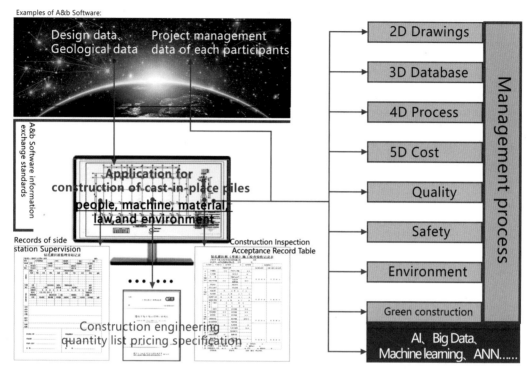

Figure 3-5 A&b Coding applications for the construction of cast-in-place piles

A&b unit is the smallest unit at different phases of lifecycle.It is executable as a WBS node or a unit of MBS. The Level 3 of A&b unit can be decomposed vertically but included in A&b applications, therefore, A&bCode breakdown has nothing to do with National engineering management processes, engineering technical standards, and work habits. A&b makes the process, organizational structure (division of responsibilities and functions), resource breakdown structure and cost system into an integrated mass, and it has all technical and management elements of projects.

Therefore, A&bCode represents a specific value of Equation（3-4）, which is called the construction information breakdown coding system(A&bCode).

Each A&bCode consist of four Arabic numerals and one English lowercase letter expressed as XYMMa, such as 2125a. The first digit X represents different phases of the lifecycle of a projcet, the second digit Y represents the "sub-project" decomposed by different phases of WBS, and the 3rd and 4th digit MM represents the specific tasks in the sub-project. The last letter indicates a sub-industry in the construction industry (such as construction engineering position a, road engineering position b……).

The construction industry information breakdown code is shown in Table 3-1 based on System Classification method.

AEC information breakdown coding(A&bCode)　　　　　　　　　Table 3-1

| |
| --- |
| A 建筑工程 Construction engineering |
| 　1000a 项目策划 Project schedule |
| 　2000a 项目规划 Project planning |
| 　2001a 规划和报建 Planning and project application |
| 　2002a 规划审批 Planning approval |
| 　3000a 项目设计 Project design |
| 　3001a 总图设计 Master chart general layout design |
| 　3002a 建筑设计 Architectural design |
| 　3100a 地基设计 Foundation design |
| 　3101a 岩土工程勘察 Geotechnical engineering survey |
| 　3102a 基坑工程设计 Foundation pit engineering design |
| 　3103a 地基处理设计 Foundation treatment design |
| 　…… |
| B 公路工程 Highway engineering |
| C 铁路工程 Railway engineering |
| D 港口与航道工程 Port and channel engineering |
| E 水利水电工程 Water conservancy and hydropower engineering |
| F 电力工程 Electrical engineering |
| G 矿山工程 Mine engineering |
| H 冶金工程 Metallurgical engineering |
| I 石油化工工程 Petrochemical engineering |
| J 市政公用工程 Municipal public engineering |
| K 通信工程 Communications engineering |
| L 机场工程 Airport engineering |
| M 核工程 Nuclear engineering |

Comparing with the construction industry task breakdown system in Figure 3-1, the construction industry information breakdown coding system is obtained as shown in Figure 3-6.

Figure 3-6 The logic frame of the AEC information coding system is as follows:

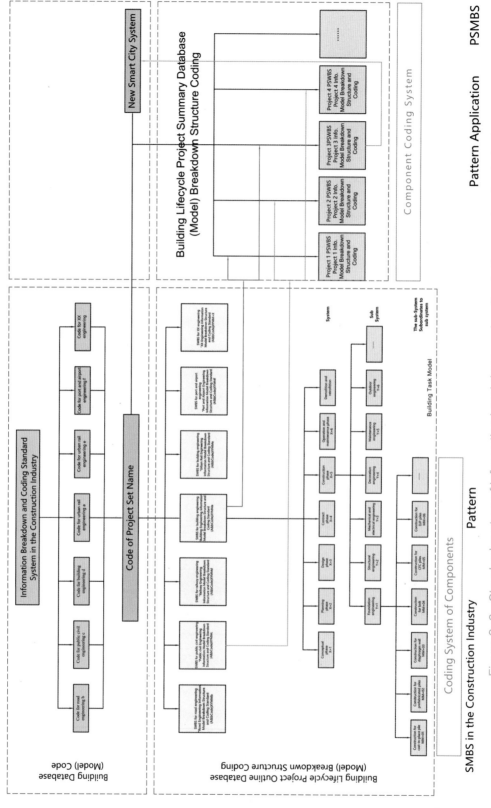

Figure 3-6　Standard system of information breakdown and coding in the construction industry

First of all, the *Uniform Standard for the Construction Industry Information Breakdown Coding* is established. The significance of this standard is to break down the whole construction industry into the coding of project sets in different sub-industries. This is the basic coding system for Digital China or Wisdom China.

Secondly, under the coding of the project set, a reusable SWBS coding system model for different sub-industries is established. Such as *Standard for breakdown and coding of highway engineering information modeling, Standard for breakdown and coding of building engineering information modeling, Standard for breakdown and coding of municipal engineering information modeling, Standard for breakdown and coding of railway engineering information modeling, Standard for breakdown and coding of airport engineering information modeling*, etc.

These two levels of breakdown coding constitute the architecture of information breakdown coding system in Chinese AEC. That is, the *Uniform Standard for AEC Information Breakdown* Coding determines the coding of all project sets in the construction industry; Then, the *XX sub-industry information model breakdown structure and coding standard*, such as *Construction Engineering School Model Breakdown Structure and Coding Standard*, determines the model breakdown model of each sub-industry project set for application to any specific project.

From the above, it can be seen that the key of information breakdown coding system in the construction industry lies in the sub-industruy information model breakdown structure and its coding standards. The practical subindustry information model breakdown structure should start with SWBS, form SMBS, and be equipped with modeling applications and information exchange standards.

A&bCode is based on the result of logical thinking, which enables BIM to be modeled.

## 3.9  A&bCode Core's Ideas and Significances

The core idea of A&bCode is to start with the end and put people first. No matter which logic way (hierarchical classification, faceted classification, mixed classification, system classification, clustered classification) is used to classify the construction industry information, the minimum value classified to the end, is that the bottom level will point to the same thing——people (users) (as shown in Figure 3-7). People are the biggest factors that drive the construction industry and constrain the development of the construction industry. Therefore, information classification should serve people and should conform to the user's logical thinking.

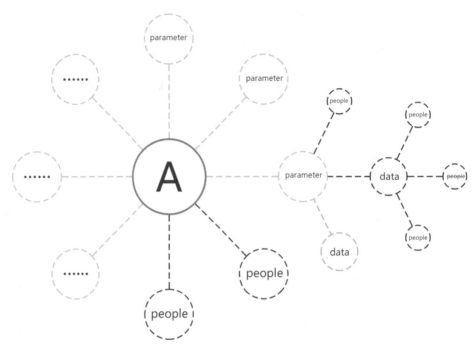

Figure 3-7  Person is always the most fundamental factor

A&bCodeis not only suitable for the informatization of engineering construction in China, but also can be internationalized. The reason is that it is the performance and definition of business flow, business flow can be additional defined by different types of operations in various countries. A&bCode is the least granular visual division of business flows, integrated and inclusive with OmniClass table 23, A&bCode is responsible for the process OmniClass table 23 is responsible for the results. A&bCode is a construction industry information drive so that BIM does not rely on the innovative development of "BIM modeling applications."

Based on MBS, the advantages of A&bCode classification coding are:

1) A&bCode describes the elements of the construction industry system engineering (system of systems) abstractly.

2) A&bCode representatively describes the work content of the deliverable entity that meets the requirements of engineering lifecycle management.

3) A&bCode is human-related, and each project participant's task corresponds to the A&bCode unit.

4) A&b applications (P-BIM functional applications) have built-in technology, management process, cost and final component products that can be delivered from different countries. Therefore A&bCode encoding has nothing to do with the country.

5) A&bCode only encodes WBS working node units. The third-level work nodes identified by different countries are essentially the same. If there are differences, they can be added.

6) A&bCode only codes the application function of the work node, it does not encode the program content and interfere with the habit of the node practitioner.

7) A&bCode is the top-level design of BIM system, which does not affect application developers' functional application innovation.

8) A&bCode aims at data exchange, which is beneficial for application developers to increase the value of functional application.

9) A&bCode creates a collaborative working environment without additional conditions for all participants involved in the project.

## 3.10 A&bCode's Demonstration of Coding Standards (Highway Engineering)

The lifecycle of a highway project is divided into six phases: planning, programming, investigation and design, contract (bidding), implementation (completion), and operations, according to the order of implementation of the project. The work and management tasks in each phase are relatively independent. And there is an upstream and downstream relationship between the project cycle chain. The division of WBS at each phase is based on the principle of construction engineering entities (buildings), and the "building" differs in its performance at different phases. It can be construction drawings, construction contracts, tunnel sub-projects, and so on. But it is all done to complete the entity of the project.

### 3.10.1 Highway Engineering SWBS Breakdown Method

The breakdown structure of highway engineering work should be based on the principle of construction engineering entities, and WBS should be carried out according to the tasks and actual work contents of different phases. For example, the design phase is first decomposed into: route design, roadbed pavement design, bridge design, tunnel design, mechanical and electrical design, house construction design, environmental protection design, and traffic safety design and etc; The contract phase is exactly the same as the implementation(completion) phase WBS. It is first decomposed into: temporary engineering, subgrade engineering, road surface engineering, bridge engineering, tunnel engineering, safety facilities and pre-buried pipeline engineering, greening and environmental protection facilities, construction engineering, electromechanical engineering. Then according to the existing construction technology, combined with the bill of quantities, quota and quality inspection and evaluation standards and other norms and standards to decompose, the subgrade engineering is divided into: earthwork roadbed, Stone roadbed, retaining wall, drainage ditch. In the lifecycle of highway construction, the principle of working breakdown structure in the same state should be unique, and WBS should not exceed 3 levels at all phases.

### 3.10.2 Highway Engineering A&b Code Encoding

The lifecycle coding of highway engineering is based on the results of WBS at each phase, using four digits + one Lowercase letters (b for highway engineering). Among them, the four digits are the first from the left, representing the division of the implementation phase of the construction project. The coding meaning is as follows: "1" represents the planningphase, "2" represents the programming phase, "3" represents the design phase, "4" represents the contract phase; "5" represents the implementation of the completion phase; "6" stands for the operations phase; "0" is designed to provide identification numbers for all relevant party management. Among the four digits, the second from the left represents the professional or work surface of the construction project. The code meaning should be determined by the actual situation in each engineering field. The second code meaning of the highway engineering is as follows: "1" represents temporary engineering. "2" represents the subgrade project, "3" represents the road surface project, "4" represents the bridge culvert project,"5" represents the tunnel project,"6" represents safety facilities and pre-buried pipelines, "7" represents greening and environmental protection facilities, "8" represents construction projects, with reference to construction projects, and "9" represents electromechanical engineering. Among the four digits, the third or fourth from the left represents the specific work content of the construction project under its profession or work surface. The code meaning should be determined by the actual work content of each engineering field. The task that defines each specific task has a unique encoding corresponding to it. The classification codes for information of every road project A&b may refer to the "*Uniform Standards for the Application of Highway Engineering Information Models*" of the Ministry of Transport, and will be replaced when there are other more relevant classification codes.

### 3.10.3  Highway Engineering WBS's Compilation Basis

(1) According to the existing construction technology, combined with the project quantity list, quotas, quality inspection and evaluation standards and other norms and standards.

(2) Cost information is compiled according to the WBS corresponding cost information at each phase. The planning phase corresponds to the *Initial (cost estimation) quotas for highway projects*, the design phase corresponds to the *Budget estimate quota for highway projects* and the *Budget quota for highway projects*, and the bidding phase corresponds to the *Budget quota for highway projects* and the *Standard construction tender document* (bill of quantities) *for highway projects*. The construction phase corresponds to the *Budget quota for highway projects* and *Construction quota for highway projects*, and the operations phase corresponds to the *Maintenance quota for highway projects*.

(3) The quality information shall be the item number of Chapter seven *Technical Specification of Highway Engineering Standard Construction Bidding Document*, the relevant requirements and highway engineering quality inspection form in the *Highway Engineering Quality Inspection and Assessment Standard* and etc.

(4) The information of the responsible person shall be compiled according to the post name and relevant regulations of the actual participant in the industry

### 3.10.4  Examples of Code Coding Standards for Highway Engineering A&bCode

As shown in Figures 3-8, Figures 3-9.

Highway Project A&b Code tender(contract) phase WBS preparation sample form

| 序号 | Subgrade Project WBS (three tiers) | | | A&bCode for lifecycle | Unified coding for Highway Engineering information modeling | Construction process | Cost information | | | Quality information | | safety Information | Information on responsible persons |
|---|---|---|---|---|---|---|---|---|---|---|---|---|---|
| | | | | | | | Budget quota number | Bill of quantities | Volume List Measurement Rules | The corresponding terms of the solicitation documents(technical specifications) | | | |
| 1 | 2 | 3 | 4 | 5 | 6 | 7 | 8 | 9 | 10 | 11 | | 13 | 14 |
| 1 | Subgrade Project | Subgrade earth and stone | | 4201B | | | | | | | | | |
| 2 | | ...... | | 4202B | | | | | | | | | |
| 3 | | Drainage Engineering | | | | | drainage ditch artificially | 1-2-1 | 203-1-a | section 207 table 207 | 207.04.3 | | | Tenderer, Bidder, Tender Agent |
| 4 | | | Drainage ditch | 4203B | | | drainage ditch of stone | 1-2-3 | 207-2-a, 207-2-b, 207-2-f | section 207 table 207 | 207.04.3 | | | Tenderer, Bidder, Tender Agent |
| 5 | | | | | | | drainage ditch of concrete | | | | | | | |
| Application positioning | collaborative applications & project management platform for lifecycle | collaborative applications for unit project | applications for tender(contract) | | | | | | | | | | |
| 6 | ...... | ...... | ...... | ...... | | | | | | | | | |

Figure 3-8   Highway Project A&bCode tender (contract) phase WBS work form

Highway Project A&b Code implementation(completion) phase WBS preparation work sample table

| No. | Subgrade Project WBS (three tiers) | | | A&bCode for lifecycle | Unified coding for Highway Engineering information modeling | Construction process | Cost information | | | Quality information | | | safety Information | Information on responsible persons |
|---|---|---|---|---|---|---|---|---|---|---|---|---|---|---|
| | | | | | | | Bill of quantities No. | Budget quota number | Construction quota No. | The corresponding terms of the solicitation documents(technical specifications) | Corresponding entry for quality inspection assessment criteria | Quality Inspection Acceptance Assessment Corresponding Form | | |
| 1 | 2 | 3 | 4 | 5 | 6 | 7 | 8 | 9 | 10 | 11 | 12 | 13 | 14 | 15 |
| 1 | Subgrade Project | Subgrade earth and stone | | 5201b | | | | | | | | | | |
| 2 | | ...... | | 5202b | | | | | | | | | | |
| 3 | | Drainage Engineering | Drainage ditch | 5203b | | | drainage ditch artificially | 203-1-a | 1-2-1 | 207.04.3 | 5.5.1 5.5.2 5.5.3 | | | Project manager, Chief Engineer, Cost Engineer, Subgrade Engineer, Builder, Quality inspector, Testers, Security officer |
| 4 | | | | | | | drainage ditch of stone | 207-2-a, 207-2-b, 207-2-f | 1-2-3 | 207.04.3 | 5.6.1 5.6.2 5.6.3 | | | Project manager, Chief Engineer, Cost Engineer, Subgrade Engineer, Builder, Quality inspector, Testers, Security officer |
| 5 | | | | | | | drainage ditch of concrete | 207-2-c, 207-2-d, 207-2-e | 1-2-4 | | | | | Project manager, Chief Engineer, Cost Engineer, Subgrade Engineer, Builder, Quality inspector, Testers, Security officer |
| Application positioning | collaborative applications & project management platform for lifecycle | collaborative applications for unit project | applications for construction | | | | | | | | | | | |
| 6 | ...... | ...... | ...... | ...... | | | | | | | | | | |

Figure 3-9   Highway Project A&bCode implementation (completion) phase WBS work form

### 3.10.5 Highway Engineering Management System

（1）P-BIM functional application

P-BIM functional application is an application function that meets the specific work needs of WBS at all phases of the engineering lifecycle. The main functions are based on different phases of business requirements including but not limited to database management, information management, progress control, cost control, quality control, safety management, and environmental protection as well as civilized construction, labor subcontract contract management, personnel management, material management, equipment management, change management , construction simulation, etc..

A&bCode coding is also the coding of application functions. The functions and data interoperability of the application shall be able to meet the requirements of collaborative work and information sharing among all relevant parties. The application shall have the functions of data import, professional inspection, achievement delivery and data delivery, and meet the requirements of data openness.

( 2 ) Collaborative application

The purpose of the collaborative application is to achieve data exchange between various phases and different functional application of each sub-project, as well as the various phases of functional application and the owner's project management platform application, the professional subcontractor's project management platform application. Data exchange between project parties management platform application to generate operational and management databases for different subprojects or various phases.

( 3 ) Highway engineering management system

The highway engineering management system includes the management platform of all parties throughout the lifecycle of projects. Management systems and platforms for all relevant units, such as transportation competent department, safety supervision department, cost management department, land management department, water Conservancy management department, environmental protection management department, fire control management department, audit department, owner unit, design unit, construction unit, supervision unit, engineering consulting unit, etc., need to be supported by functional applications and collaborative applications.

# Chapter 4　HIM Interoperability Based on A&bCode

## 4.1　Digital Thread Network

### 4.1.1　Open CNC System

The IEEE (American Institute of Electrical and Electronics Engineers) definition of open systems is: It can run on multiple platforms, interoperate with other systems, and provide users with a unified style of interaction. Generally speaking, the open CNC system allows users to choose and integrate according to their own needs, change or expand the functionality of the system to quickly adapt to different application requirement, and the functional modules of the system can be sourced from different component suppliers and be compatible with each other.

（1）US GNC program

The United States is the initiator of the open CNC system. In 1987, NGC (Next Generation Workstation/Machine Controller) program was proposed.The goal of the NGC project is to provide a standard for the next generation of manufacturing controllers based on open system architectures that allow different designers to develop interchangeable and interoperable controller components. For example, coordination between multiple devices, fully independent programming of devices, model-based processing, adaptive path strategies, and a wide range of workstations and real-time features, etc.For example, coordination among multiple devices, fully independent programming of devices, model-based processing, adaptive path strategy and a wide range of workstations and real-time characteristics, etc.The architecture of NGC is established on the basis of virtual machine, which links the system and module to the computer platform.

(2) European OSACA program

The OSACA program was jointly developed in 1990 by 22 controller developers, machine tool manufacturers, control system integrators and research institutes in the European Community.The architecture of "layered system platform + structured functional unit" is proposed by the OSACA program.The architecture ensures the independence of various application systems and operating platforms and the interoperability between them, ensuring openness.

(3) Japan's OSEC program

Japan's OSEC program is jointly established by three machine tool manufacturers, Toshiba Machine Company, Toyota Machine Works and Mazak's, and Japan's IBM, Mitsubishi Electric and SML Information Systems.The aim is to establish an international standard for factory automation control equipment.In terms of hardware, OSEC plans to adopt the structure of PC + control card, which is conducive to the realization of hierarchical, modular and flexible configuration.OSEC groups and structures functional units in some functional layers, and its open architecture consists of three functional layers with a total of seven processing levels.

Compared with the international advanced level, the research of open CNC system in China is still in its infancy.

### 4.1.2 Digital Thread

The Digital Twin describes the model of each specific link connected through the Digital Thread.Digital Thread establishes a technological process through advanced modeling and simulation tools to provide the ability to access, synthesize and analyze data at all phases of the system's lifecycle.Based on high fidelity system model, all departments can make full use of seamless interaction and integration analysis of various technical data, information and engineering knowledge to complete real-time analysis and dynamic evaluation of project cost, schedule, performance and risk.Digital Thread has the characteristics of "all element modeling definition, all data acquisition and analysis, all decision simulation evaluation".It can quantify and reduce various uncertainties in the lifecycle of the system, and realize the automatic tracking of demand, rapid iteration of design, stable control of production and real-time management of maintenance.Throughout the lifecycle, the models of each link can synchronize and communicate key data in two directions in time.Based on these models, the current and future functions and performance of the system can be dynamically and real-time evaluated.

Digital Thread runs through the product lifecycle, especially seamless integration from product design, production, operation and maintenance.Digital Twin is more like the concept of intelligent products, which emphasizes the feedback from product operation and maintenance to product design.

### 4.1.3 Digital Thread Network

Although Omniclass is a scientific classification coding system, it is not perfect and there are many problems in its application.Omniclass tries to create an omniscient God, to classify all the attributes.It wants users to pick it out on their own.As a result, the coding system is so complicated and can not cover all the classifications that it can't be applied in practice.

When the road is blocked, we need to find out why.Maybe it's the builder's problem, or maybe it's the direction of the building.With so many coding organizations and so much manpower, the problem has not been solved yet.Using converse thinking to explore what we really want, it is a structured integrated database with integrated information, delivery on demand, collaborative creation and open standards.Constructing structured database of engineering entity and process business in construction industry is the information integration of engineering and business and the digital expression of engineering and business.

The true face of Lushan is lost to my sight, for it is right in this mountain that I reside.If we change our thinking and direction, it will be another world.Information Breakdown Coding in Construction Industry is an information breakdown system that classifies the information of construction industry (construction industry) and applies it directly to the informatization of construction projects.A&bCode is a summary of the logical structure of OmniClass.

For construction projects, the project is composed of many different products.Not only should digital twin be seamlessly integrated from product design, production, operation and maintenance, but also different products should be co-constructed.Therefore, Digital Thread in manufacturing industry must be Digital Thread Network in construction industry.

HIM (A Digital thread Network) integrates the concepts of open CNC system and Digital Thread, which based on A&bCode.Then, the interoperability of BIM will be realized as building the Digital Thread Network. It is a collaborative platform for data exchange in construction industry similar to the concept of open CNC system.

## 4.2 HIM Digital Thread Network Based on A&bCode

### 4.2.1 Force Method Equation of n-order Statically Indetermine Structure

For n-order statically indeterminate structures, the basic structure of force method can be obtained by removing $n$ redundant constraints in force method calculation. Replacing $n$ redundant constraints with multiple constraints combined with the original load to form the basic system of force method. When the displacement of the original structural system that the redundant constraints are removed is zero, there are corresponding n known displacement conditions to establish $n$ equations for solving the redundant constraints:

$$\begin{aligned}
\delta_{11}X_1 + \delta_{12}X_2 + \cdots + \delta_{1n}X_n + \Delta_{1P} &= 0 \\
\delta_{21}X_1 + \delta_{22}X_2 + \cdots + \delta_{2n}X_n + \Delta_{2P} &= 0 \\
&\cdots\cdots \\
\delta_{n1}X_1 + \delta_{n2}X_2 + \cdots + \delta_{nn}X_n + \Delta_{nP} &= 0
\end{aligned} \qquad (4\text{-}1)$$

Equation (4-1) can be written as a matrix expression:

$$\begin{bmatrix} \delta_{11} & \delta_{12} & \cdots & \delta_{1n} \\ \delta_{21} & \delta_{22} & \cdots & \delta_{2n} \\ \cdots & & \ddots & \cdots \\ \delta_{n1} & \delta_{n2} & \cdots & \delta_{nn} \end{bmatrix} \begin{Bmatrix} X_1 \\ X_2 \\ \cdots \\ X_n \end{Bmatrix} = \begin{Bmatrix} -\Delta_{1P} \\ -\Delta_{2P} \\ \cdots \\ -\Delta_{nP} \end{Bmatrix} \qquad (4\text{-}2)$$

The left side of Equation (4-2) becomes the flexibility coefficient matrix, and $\delta_{ij}$ is called the displacement influence coefficient or the flexibility coefficient. (Figure 4-1).

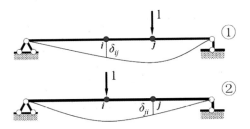

Figure 4-1  $\delta_{ij}$ equals the displacement corresponding to $F_{Pi}$ caused by $F_{Pj}=1$

Among them $\delta_{ij}$, the subscript $i$ denotes the point at which displacement occurs, and the subscript $j$ denotes the point at which unit force acts, which is the cause.

### 4.2.2 HIM Digital Thread Network Based on A&bCode

We can use A&bCode to establish HIM (the Digital thread Network of BIM) according to the same concept in Equation (4-2).

If we interpret the flexibility factor of formula (4-2) as a corresponding database that $F_{Pj} = 1$ delivered to another application $F_{Pi}$, then the matrix expression of $HIM_i$ for any working node integrated database is:

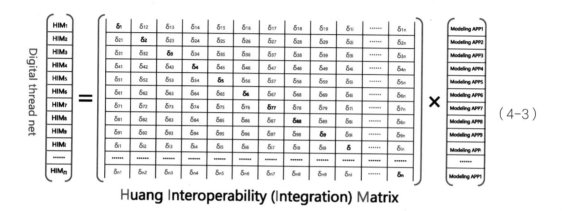

**H**uang **I**nteroperability (**I**ntegration) **M**atrix

Taking $A\&b_{ij} = \delta_{ij} \times \text{Modeling Application}_j$, then equation (4-3) can be expressed as:

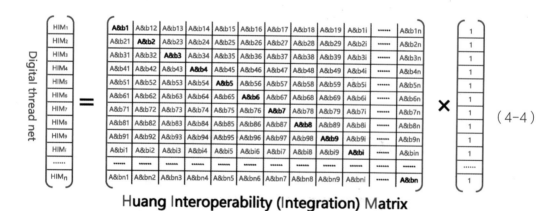

**H**uang **I**nteroperability (**I**ntegration) **M**atrix

The physical meaning of Equation (4-4) is shown in Figure 4-2.

Equation (4-4) shows the application interoperability exchange database folder. The contents and format of the exchange folder of the modeling application $i$ are determined by standards. Therefore, for different modeling applications (application functions), the corresponding information exchange standard is required, and Equation (4-4) should be further expressed as:

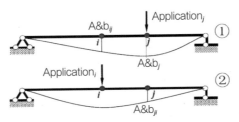

Figure 4-2  Displacement influence diagram of Application Data interoperability

$$\begin{pmatrix} HIM_1 \\ HIM_2 \\ HIM_3 \\ HIM_4 \\ HIM_5 \\ HIM_6 \\ HIM_7 \\ HIM_8 \\ HIM_9 \\ HIM_i \\ \cdots \\ HIM_n \end{pmatrix} = \begin{pmatrix} A\&b1 & A\&b12 & A\&b13 & A\&b14 & A\&b15 & A\&b16 & A\&b17 & A\&b18 & A\&b19 & A\&b1i & \cdots & A\&b1n \\ A\&b21 & A\&b2 & A\&b23 & A\&b24 & A\&b25 & A\&b26 & A\&b27 & A\&b28 & A\&b29 & A\&b2i & \cdots & A\&b2n \\ A\&b31 & A\&b32 & A\&b3 & A\&b34 & A\&b35 & A\&b36 & A\&b37 & A\&b38 & A\&b39 & A\&b3i & \cdots & A\&b3n \\ A\&b41 & A\&b42 & A\&b43 & A\&b4 & A\&b45 & A\&b46 & A\&b47 & A\&b48 & A\&b49 & A\&b4i & \cdots & A\&b4n \\ A\&b51 & A\&b52 & A\&b53 & A\&b54 & A\&b5 & A\&b56 & A\&b57 & A\&b58 & A\&b59 & A\&b5i & \cdots & A\&b5n \\ A\&b61 & A\&b62 & A\&b63 & A\&b64 & A\&b65 & A\&b6 & A\&b67 & A\&b68 & A\&b69 & A\&b6i & \cdots & A\&b6n \\ A\&b71 & A\&b72 & A\&b73 & A\&b74 & A\&b75 & A\&b76 & A\&b7 & A\&b78 & A\&b79 & A\&b7i & \cdots & A\&b7n \\ A\&b81 & A\&b82 & A\&b83 & A\&b84 & A\&b85 & A\&b86 & A\&b87 & A\&b8 & A\&b89 & A\&b8i & \cdots & A\&b8n \\ A\&b91 & A\&b92 & A\&b93 & A\&b94 & A\&b95 & A\&b96 & A\&b97 & A\&b98 & A\&b9 & A\&b9i & \cdots & A\&b9n \\ A\&bi1 & A\&bi2 & A\&bi3 & A\&bi4 & A\&bi5 & A\&bi6 & A\&bi7 & A\&bi8 & A\&bi9 & A\&bi & \cdots & A\&bin \\ \cdots & \cdots & \cdots & \cdots & \cdots & \cdots & \cdots & \cdots & \cdots & \cdots & & \cdots \\ A\&bn1 & A\&bn2 & A\&bn3 & A\&bn4 & A\&bn5 & A\&bn6 & A\&bn7 & A\&bn8 & A\&bn9 & A\&bni & \cdots & A\&bn \end{pmatrix} \times \begin{pmatrix} 1 \\ 1 \\ 1 \\ 1 \\ 1 \\ 1 \\ 1 \\ 1 \\ 1 \\ 1 \\ \cdots \\ 1 \end{pmatrix} \quad (4-5)$$

数字素网（Digital thread net）

**Huang Interoperability (Integration) Matrix**

As A&bCode integrates WBScode, MBScode, application function code and application function information exchange standard code, the actual meaning of HIM matrix is shown in Figure 4-3.

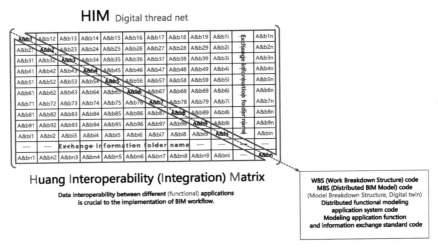

Figure 4-3  The practical meaning of HIM

The simplified expression of HIM is shown in Figure 4-4.

| A&b1 | A&b12 | A&b13 | A&b14 | A&b15 | A&b16 | A&b17 | A&b18 | A&b19 | A&b1i | …… | A&b1n |
|---|---|---|---|---|---|---|---|---|---|---|---|
| A&b21 | A&b2 | A&b23 | A&b24 | A&b25 | A&b26 | A&b27 | A&b28 | A&b29 | A&b2i | …… | A&b2n |
| A&b31 | A&b32 | A&b3 | A&b34 | A&b35 | A&b36 | A&b37 | A&b38 | A&b39 | A&b3i | …… | A&b3n |
| A&b41 | A&b42 | A&b43 | A&b4 | A&b45 | A&b46 | A&b47 | A&b48 | A&b49 | A&b4i | …… | A&b4n |
| A&b51 | A&b52 | A&b53 | A&b54 | A&b5 | A&b56 | A&b57 | A&b58 | A&b59 | A&b5i | …… | A&b5n |
| A&b61 | A&b62 | A&b63 | A&b64 | A&b65 | A&b6 | A&b67 | A&b68 | A&b69 | A&b6i | …… | A&b6n |
| A&b71 | A&b72 | A&b73 | A&b74 | A&b75 | A&b76 | A&b7 | A&b78 | A&b79 | A&b7i | …… | A&b7n |
| A&b81 | A&b82 | A&b83 | A&b84 | A&b85 | A&b86 | A&b87 | A&b8 | A&b89 | A&b8i | …… | A&b8n |
| A&b91 | A&b92 | A&b93 | A&b94 | A&b95 | A&b96 | A&b97 | A&b98 | A&b9 | A&b9i | …… | A&b9n |
| A&bi1 | A&bi2 | A&bi3 | A&bi4 | A&bi5 | A&bi6 | A&bi7 | A&bi8 | A&bi9 | A&bi | …… | A&bin |
| …… | …… | …… | …… | …… | …… | …… | …… | …… | …… | …… | …… |
| A&bn1 | A&bn2 | A&bn3 | A&bn4 | A&bn5 | A&bn6 | A&bn7 | A&bn8 | A&bn9 | A&bni | …… | A&bn |

Figure 4-4　HIM simplified expression

Looking back at the 20 years of BIM development since the first edition of IFC, the development of BIM is the time to move from "component thinking" to "model thinking" and from "Highbrow BIM" to "Popular BIM". Many experts often say, "Application defines the world and data drives the future", and so does BIM. BIM is a system composed of many independent application subsystems. Data can drive application, and application can produce data. If the data is food, then application is a tool for processing grain into food. Only when the two are combined will our world change and we welcome a feast of BIM industry.

## 4.3　HIM for Compatibility and Interoperability

The basic opinion of structuralist anthropology are as follows:The external world we know is learned through consciousness. The phenomena we observe have the characteristics we give them. This is determined by the way we operate and the way the human brain organizes and interprets the stimuli.A very important feature of this sorting process is that we can cut the time and space around us into continuous pieces, so that we can first regard the environment as a large number of individuals that can be attributed to a certain name. The composition of things, the time is seen as a sequence of isolated time.Similarly, as human beings, when we make artificial products (various artifacts), such as preparing for a ceremony or writing past history, we will imitate our understanding of nature;Thus, as we have determined that natural things are divided and ordered, we divide and arrange cultural products (Leach, 1970).This paragraph is a simple interpretation of Levi Strauss's structural anthropology.

Different people may have different foresights for future construction.But when it comes to the future of the construction process, everyone will agree with the same development trend:automation and intelligence.From the key graphs of Grasshopper, we found that each task in each phase is a data-supported and logical organization.If we put all the processes in one diagram, and the logical lines of the different processes between each object are represented by different colors at the same time, we will see that our entire construction process is a network (Figure 4-5). Each node in the network is each task involved in the construction.

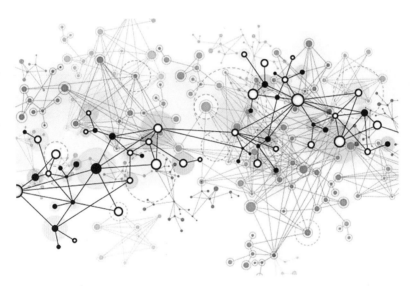

Figure 4-5　Future construction process

Topologies represent the positions of various objects as abstract positions.In the network, the topology visually describes the arrangement and configuration of the network, including the relationship between various nodes and nodes. Topology does not care about the details of things, nor does it care about the proportional relationship between them. It only expresses the relationship between things in the scope of discussion, and shows the relationship between these things through graphs.

The mathematical expression of HIM as a topology map for future construction is shown in Figure 4-6. Any construction node in the left diagram can be orderly split through HIM on the right (Figure 4-7).

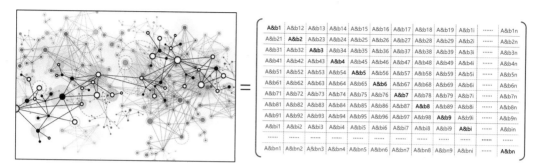

Figure 4-6   HIM expression of the topology

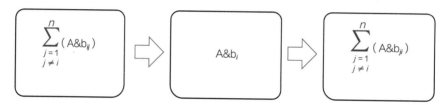

Figure 4-7   HIM deconstruction of complex construction process

Thus, a smart construction mode based on the HIM network operating system is formed (Figure 4-8).

A&bCode is equivalent to a fixed IP address in the network, which can provide a stable logical relationship to make engineering logic into data exchange logic. Pre-implantation into the network exchange rules established by HIM enables the data transmitted from the starting end to the demand end through A&bCode, and the information can be precisely customized (Figure 4-9).

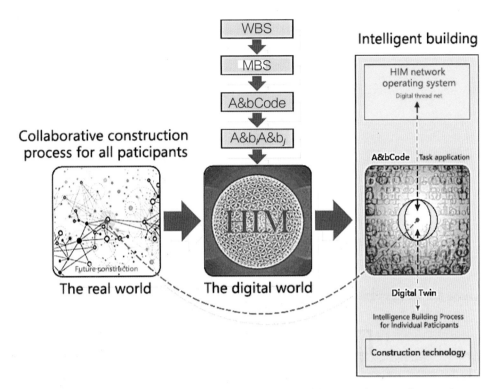

Figure 4-8  Smart construction mode based on HIM network operating system

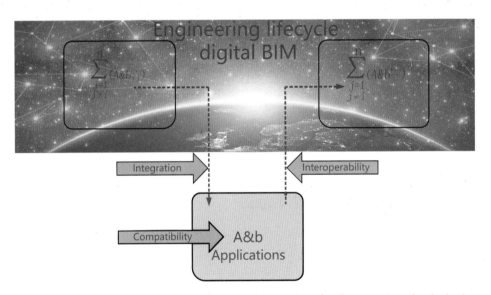

Figure 4-9  Peer-to-peer network operating system for the construction industry

# Chapter 5   BIM with A&bCode

## 5.1   BIM Standard Systems of the United States and China

The National BIM Standard (NBIMS) system is shown in Figure 5-1。

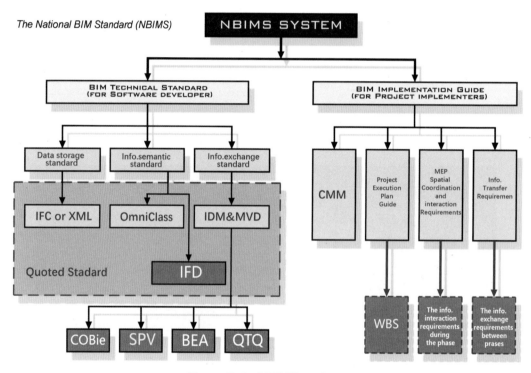

Figure 5-1   NBIMS system

China P-BIM standard system is shown in Figure 5-2。

It can be seen from the comparison between Figure 5-1 and Figure5-2 that the difference between the U.S. BIM standard system and China P-BIM standard system lies only in the different cognition of decision-makers who lead the application of BIM technology. Different cognition leads to different BIM implementation methods. China's

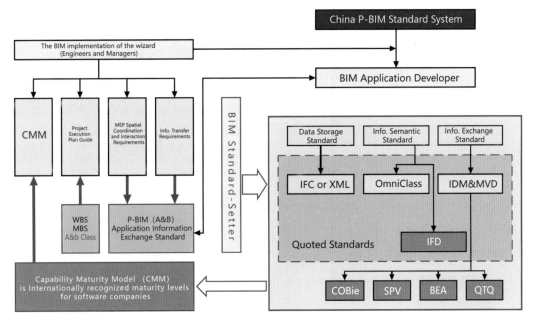

Figure 5-2　P-BIM system

National Unified Standard for Application of Building Information Models contains two different BIM implementations (Figure 5-3)。

From Figure 5-3, it can be seen that the international mainstream BIM standard system adopts the construction information classification standard represented by OmniClass, while the P-BIM standard system adopts the construction information breakdown standard represented by A&bCode to implement BIM. Therefore, the P-BIM standard system is compatible with the international mainstream BIM standard system.

## 5.2　A&bCode and OmniClass

Compare the A&bCode coding with the 15 tables of OmniClass. The contents of the 15 tables of OmniClass15 can correspond to the A&bCode coding respectively (Figure 5-4).

## 5.3　Information Exchange Architecture between A&bCode and IFC, NBIMS

It is well acknowledged that the IFC outline is layered and modular (the overall framework). Shown in Figure 5-5

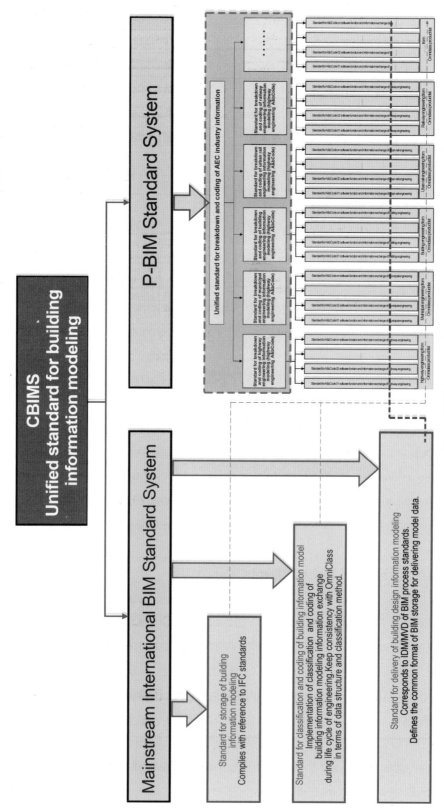

Figure 5-3 China BIM Standard system

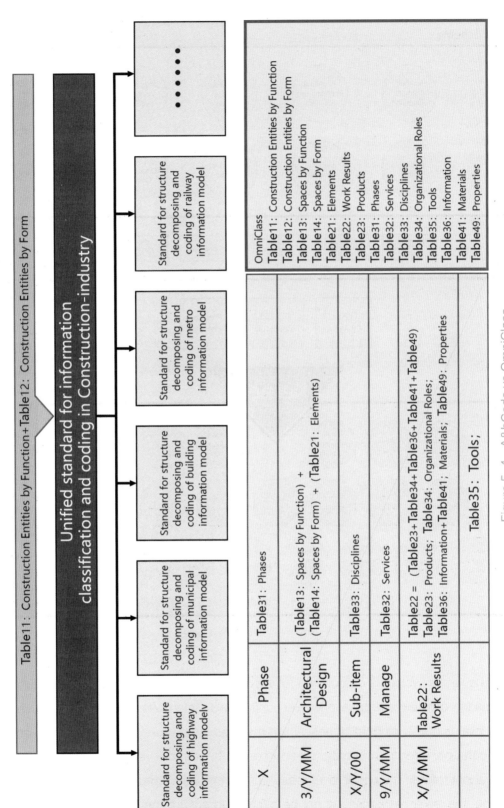

Figure 5-4 A&bCode vs OmniClass

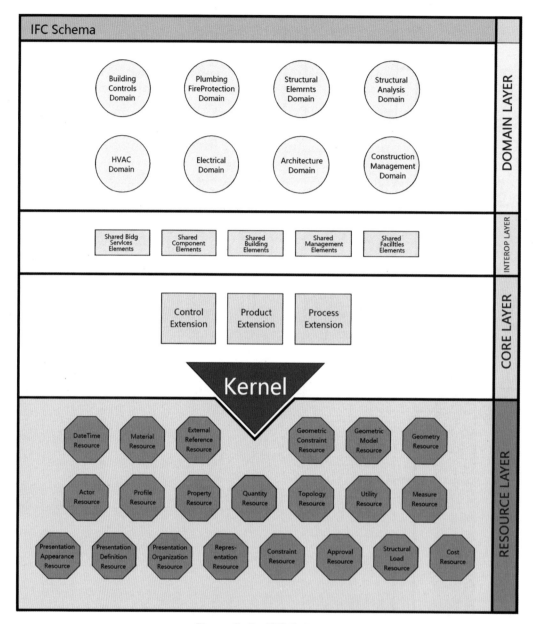

Figure 5-5  IFC Schema

The whole system is divided into four levels, including information resource layer, information core layer (framework layer), information interop layer and domain layer. Each level contains information description modules and follows the principle that each level can only refer to information resources of the same level and the lower level, and cannot refer to the upper level resources. In this way, when the upper resources change, the lower resources will not be affected to ensure the stability of information description. Contents of each level include:

1) The information resource layer describes the basic information used in the standard, such as personnel information, document information, geometric topology information, etc. These basic information are not for construction engineering and equipment management, but only scattered information without overall structure. They will be used as the basis of information description and applied to the entire information model.

2) The information framework layer (core layer) describes the overall framework of architectural engineering information. It organizes the information of the resource layer with a regular framework and makes them interrelated and connected to form a whole, truly reflecting the structure of the real world.

3) The information sharing layer solves the problem of domain information interaction. At this level, each system refines the constituent elements and ensures that there are no missing elements.

4) The domain layer penetrates into the interior of each field to form thematic information in various fields.

There are several modules in each level.

Data of IFC building design has a strong specialty. In the IFC2x2 final version of the IFC pattern, 623 entities, 110 specific types, 159 enumerated types, and 42 selection types are defined, and there are complicated relationships between the entities. It is the instances of these entities and the relationships between them that form the complete IFC building model.

For ordinary application developers do not need to understand the full content of the IFC standard, only need to understand the corresponding part in the case of clear overall framework and core structure.

In the AEC/FM industry, IFC has more and more scientific research and practical application projects. Under such circumstances, people find that the way of transmitting information through documents is not conducive to the development of the industry. Because in the application, people often only need part of the information in the file, and only the entire model can be transmitted through the file transfer, which greatly increases the system overhead, and the data is not easy to use. Therefore, a data sharing system for IFC local model data has emerged. Since then, local model data exchange has been a hot issue in the related research of IFC standards, but it has not formed a general, flexible and highly scalable method to describe IFC local model data query information and implement IFC. Local model data exchange control.

Due to the breadth of application of the IFC standard, there are various query description methods that are provided to the user to describe the required local model data (or client). For example, using a graphical user interface, some special values can be set to select a local model, such as an object with a door height of no more than 200cm. At the same time, when different kinds of client programs are connected to the IFC local model server, they may have different needs for the selection of local models. For example, client A requires specific building element objects and related geometric objects. What client B needs to select the same building element object and its associated cost information object. Therefore, the IFC local model sharing server must be able to adapt to different query situations described above and give corresponding local model data.

IFC standard promoters have been looking for a programmable query language to support the manipulation of local model data, which can satisfy the client terminal's query of various description methods and contents, and avoid the difficulty of user interaction. Only by using a program-independent local model query language can the processing algorithm of the query description be written directly in different programs, thus ensuring the simplicity, flexibility and scalability of the system. However, the local model requirements in the whole life of the construction project are ever-changing, and the local model requirements are difficult to determine. This query language cannot be written.

The domain layer of the project only contains physical and domain services. The entity is "building", including different "construction" methods of design and construction process, as shown in Figure 5-5, "HVAC", "Electrical", "construction"; the field service behavior is the "building-management", such as "construction management" and "property management".

The construction project is a complex, integrated multi-disciplinary activity in which professional activities are domain entities; the behavior of management domain entities is domain services. In other words, the fourth "domain layer" in Figure 5-5 should contain all the independent "activities or businesses" of the lifecycle (Figure 1-4, U.S. NBIMS data exchange hierarchy diagram).

No application developer can independently provide an application that covers the entire lifecycle of a building, and no project can be completed using only one application company's application products. Therefore, solving the problem of information exchange and sharing is very important for the construction industry. Current building application is only involved in a specific field application at a certain phase of the construction engineering lifecycle. The IFC Standard Domain Layer (IFC-Domain Layer), as the top level of the IFC architecture, defines the types of entities for each area of expertise. These entities have specific concepts for each area of expertise. In order for IFC to meet the information exchange requirements of the lifecycle, it is necessary to define all A&bs of the domain layer, namely A&bCode.

A domain model is a visual representation of a conceptual class in the domain or an object in the real world. The domain model, also known as the conceptual model, the domain object model, and the analytical object model, is dedicated to interpreting important things and products in the business domain. The domain layer is a conceptual model in a three-tier database structure (Figure 5-6)

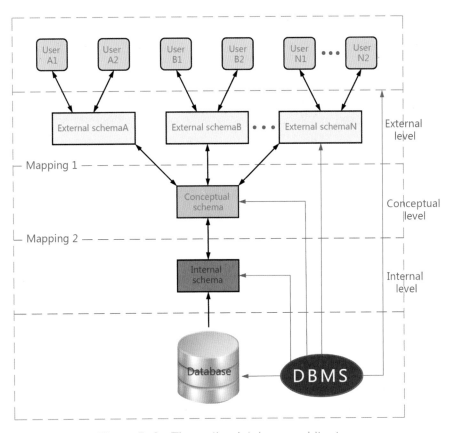

Figure 5-6   Three-tier database architecture

Without a clear definition of the domain layer, the good ideals of the IFC's overall framework will be hard to accomplish. IT project staff can hardly meet the ever-changing needs of engineering and manager even though they have a cat's whiskers. This is the task of professionals and managers outside of IT project, no professional leading domain entities and field service for the domain analysis, the IFC standard and buildingSMART and NBIMS interaction theory is hard to fall to the ground by project engineer application.

IT is only through an in-depth analysis of the IFC domain layers, listing all A&B-items throughout the life of a construction project, according to these requirements, IT project staff are likely to achieve local model interaction needs. The introduction of A&b-Code fills the gap in the IFC architecture domain layer and completes the IFC architecture. The domain layer needs to establish a domain model. The first step in building a domain model is to identify entities, value objects, and domain services.

- Entity: An entity is a domain concept that needs to be uniquely identified in the domain. Usually in business, there is a class of objects that need to be uniquely identified and distinguished, and continue to track it. Such objects are considered entities. The unique identifier here usually refers to the unique identifier of the business, such as the order number, employee number, andsoon, rather than the int increment id or Guid column stored in the database because of the technology needs; the entity only retains the necessary attributes and behaviors. For example, for a customer entity, the basic information of the customer should be retained, such as: name, gender, age, andsoon; but information such as country, province, city, street, andsoon. Jointly represents the complete subordinate concept of the customer, and this complete subordinate concept should be migrated to On other entities or value objects, this helps to understand and maintain the client entity and clearly define the objec responsibilities.

- Value object: A value object is a domain concept that does not need to be uniquely identified in the domain. In business usually, we do not need to distinguish which object is specific, but only what state the object is. Such an object is considered to be a value object. Both objects have the same state, we think it is the same object, such as address, order status, and so on. Value objects are read-only, cannot be modified directly with invariance, but can be replaced.

● Domain services: When certain business activities can neither be defined as an entity nor attributed to value objects, they can be attributed to the concept of domain services. Domain services are essentially operations that do not contain state and are often used to coordinate multiple entities. Domain services can be read directly to the application layer, which effectively protects the domain model.

The domain/application layer is the highest level of the IFC architecture, providing the conceptual model needed for the construction and facility management areas. Currently, the domain models defined by IFC include Architecture, Facility Management, Cost Estimating, HVAC. This definition is at best only part of the design phase model, A&b-Code is based on WBS and MBS for construction projects and is a model for building lifecycle management.

The architecture of IFC follows the ladder principle: each level of the class can refer to other classes of the same level or lower level, but not to higher level categories. It can be seen that the A&b domain model located in the domain layer can call the classes of this layer and the lower layer. Therefore, when the "A&b Application Function and Information Exchange Standard" is formulated, the IFC standard, the partial application of the IFC standard, and the complete use of the IFC standard can be fully utilized according to various practical conditions. This not only compensates for the imperfect IFC standards for construction projects, but is also extremely important for infrastructure engineering BIM applications that do not yet have IFC standards.

Therefore, the "A&bCode Coding System Standard" is the basic standard for BIM technology innovation and implementation.

IFC is an intermediate data format. All applications for building lifecycle management is exchanged based on IFC. A&b-Code describes a collection of all application concept schema in the IFC domain layer. Without A&b-Code, even the concept schema of the IFC domain layer cannot implement the external schema. The NBIMS information exchange architecture is to develop an external schema for different A&b-Code (Figure 1-4). Therefore, the lack of A&b-Code, the NBIMS is also hard to accomplish.

Figure 5-7 shows the NBIMS, the IFC outline (Figure 5-5), the database three-tier structure (Figure 5-6), and the A&b-Code and relationship connections。

Figure 1-4 is associated with Figure 5-7 of the information exchange architecture to obtain Figure 5-8。

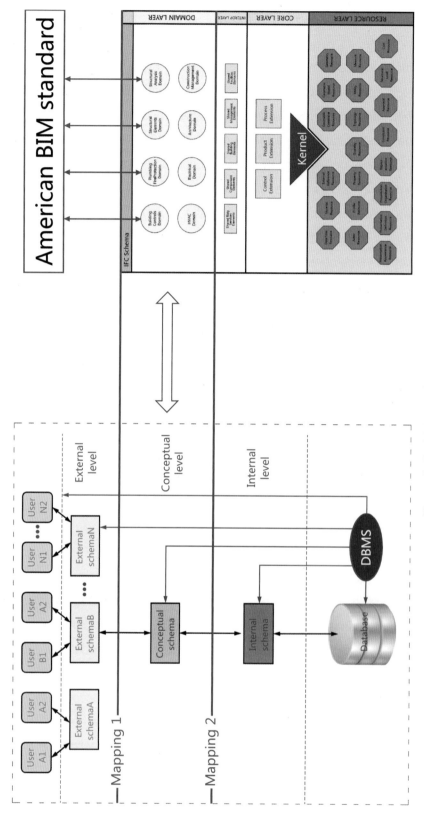

Figure 5-7 Main work of the NBIMS

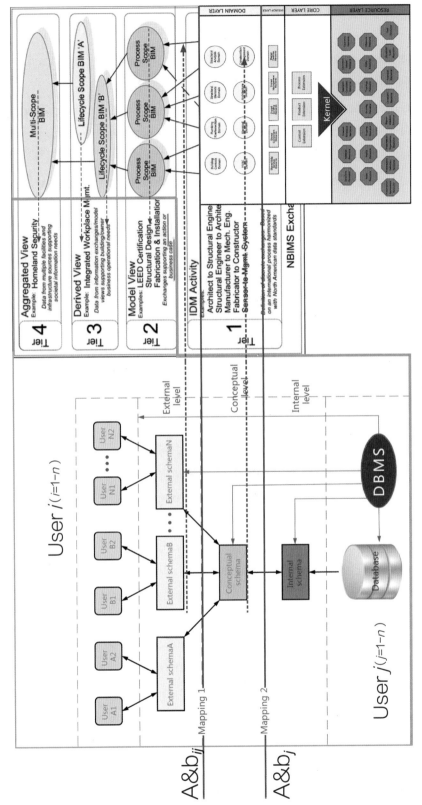

Figure 5-8 A&b-Code, NBIMS Info.Exchange Architecture, IFC, user relationship diagram

Chapter 5 BIM with A&bCode

## 5.4 A&bCode and IDM/MVD

The A&bCode-based HIM and NBIMS information exchange architecture, IFC, and user relationships are shown in Figure 5-9。

Figure 5-9 shows the corresponding MVD expression of the A&bCode applications exchanged by IDM directly in different roles, that is, IDM and MVD are completely mapped. A&bCode A&b$_{ij}$ has the same concept as IDM/MVD.

A&bCode refines the domain layer of IFC and the application layer in the NBIMS information exchange architecture (Figure 5-10)。

The difference between the P-BIM standard recommended by China national standard *Unified Standard for the Application of BIM*（GB/T 51212—2016）and the implementation of the NBIMS is shown in Figure 5-11.

It can be seen from Figure 5-11 that when the data between the application in the P-BIM standard is not exchanged, it is the NBIMS standard pattern on the right side of Figure 5-11, but the collaborative application becomes the modeling for the MVD.

The series of standards *P-BIM Application Functions and Information Exchange Standards* prepared by the China BIM Development Alliance has completed some of the standards for the construction engineering design phase (Figure 5-12).

As can be seen from Figure 5-9, the HIM-based BIM implementation directly converts the IDM of the human language exchange mode shown in Figure 5-13 to the interoperability between different applications of the computer through A&bCode (Figure 5-14). This method is to start with the finalization: directly compile the MVD standard, and the application company completes the interface program to implement BIM.

The China BIM Development Alliance is based on Figure 5-14 the A&b compatible application interoperability (HIM-based Engineering lifecycle digital BIM platform) laboratory has been built and used in Shenzhen University (Figure 5-15).

The construction information classification A&bCode based on WBS and MBS is closely integrated with engineering practice (through the building lifecycle management) and closely related to the employees involved in the project (people-oriented), so that BIM implementation and engineering tasks are integrated. By developing the application of each part of the project (A&b), it is a simple and demand-oriented BIM implementation method to determine the contract and design phase data exchange standards. It provides a huge space for China's BIM innovation and entrepreneurship

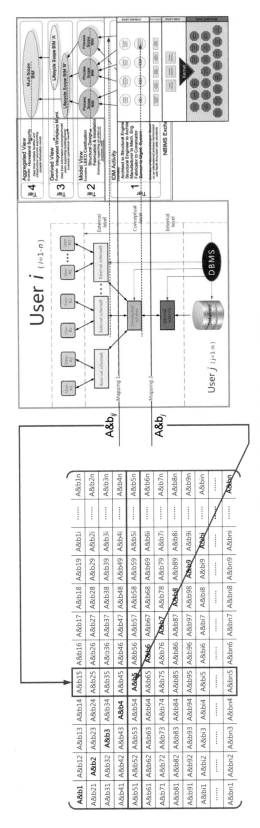

Figure 5-9 A&bCode-basedHIM

Chapter 5 BIM with A&bCode 167

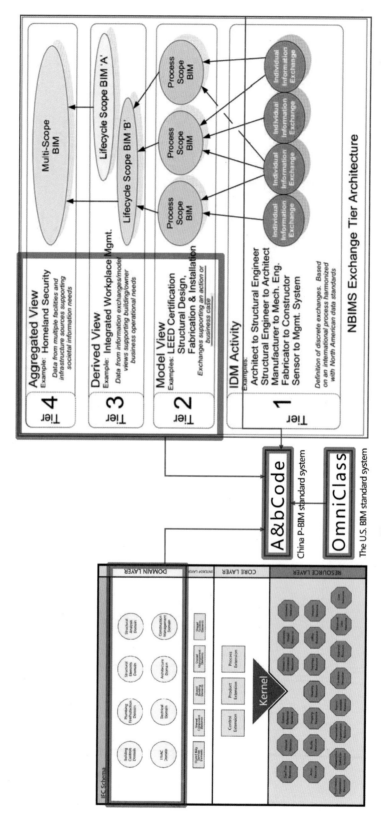

Figure 5-10 The significance of A&bCode

(rather than modeling). The information breakdown method determined by the breakdown structure of building information model is of great significance to the development of BIM technology innovation in China.

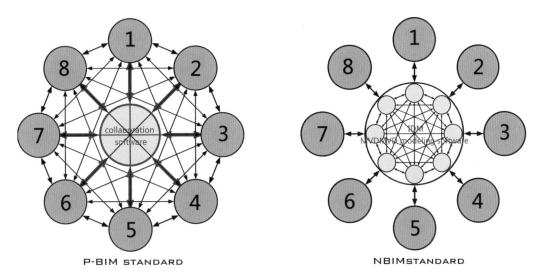

1 APPLICATIONS FOR ARCHITECT
2 APPLICATIONS FOR STRUCTURAL ENGINEER
3 APPLICATIONS FOR HVAC ENGINEER
4 APPLICATIONS FOR CONTROL ENGINEER
5 APPLICATIONS FOR CONSTRUCTION MANAGER
6 APPLICATIONS FOR FACILITY MANAGER
7 APPLICATIONS FOR OWNER
8 APPLICATIONS FOR CIVIL ENGINEER

Figure 5-11  P-BIM vs NBIMS

Figure 5-12  Standard for P-BIM Application Functions and Information Exchange

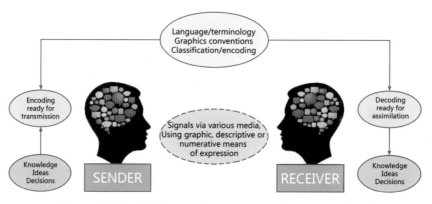

Figure 5-13　IDM human language exchange standard

Figure 5-14　A&b-based Application interoperability

Figure 5-15　HIM-based internet platform for the construction industry

# Acknowledgements

I have been involved in BIM study since 2012. For seven years, I have spent most of my time and efforts on it. Taking this opportunity, I would very much like to extend my special gratitude to Xu Jiefeng, Zhu Lei, Jin Xinyang, Lou Yueqing, Jin Rui, Wu Jun, Huang Kun, Liu Hongzhou, Li Shaojian, Gong Jian, Li Yungui, Zhang Jianping, Xu Jianzhong, Zuo Jiang, Wang Xiaojun, Cheng Zhijun, He Guanpei, Gan Jiaheng, Mao Zhibing, Wang Dan, Xie Wei, Gao Chengyong, Wang Jiayuan, Li Dongbin, Wang Rong, Zhang Miao, Ren Feifei, Zhang Zheng of China BIM Union for their full support and assistance offered to me.

Today, the A&bCode research group and I, as the head, are very pleased to have our research findings published. From time to time the vivid pictures and unforgettable memories come to me about my colleagues and others who have guided me, helped me, and cared about me over the past seven years. Thank you all.

In April 2012, Beijing Lizheng Application Design and Research Institute Co., Ltd. designed the concept map of P-BIM for me, and later it became the logo of China BIM Union. My colleagues and I have been working hard along this goal.

In June 2012, I participated in the BIM delegation of the Ministry of Housing and Urban-Rural Development to visit the United States. I was accompanied by Dr. Gan Jiaheng and the investigation made me know a few key points of BIM:

1. Mr. Deke Smith, buildingSMART Alliance: The biggest possible risk in the development and implementation of BIM is "overselling".

2. Mr. Mark A. Kohl, Imagineering, Florida Disney: What you get out of BIM is a result of what you put into BIM.

3. Mr. Jon Pittman, Vice President, Global Corporate Strategy, Autodesk: We are tool manufacturers, not ourselves to make the world more exciting. We are just making tools to help designers make the world more wonderful.

4. Mr. Martin Fischer, Director of the CIFE Center at Stanford University: We want to prepare the next generation of employees for the environment they are willing to work in (AEC industry).

5. Mr. Patrick MacLeamy, Chairman of the buildingSMART International Headquarters: Why are you so keen on BIM?

6. USA HOK VDC/BIM team: BIM should be convenient to use, and the data generated by different applications should be stored in the Excel table.

With these views and issues, we have been committing to establish China's P-BIM standard system since 2012. Thanks to all the units and individuals involved in the P-BIM standard development over these years.

中国BIM标准研究项目验收证书

Thanks to Shenzhen University for establishing a data exchange laboratory (Internet of AEC Industry and BIM Lab) for P-BIM application.

In November 2017, the Chairman of buildingSMART International Headquarters

Patrick MacLeamy and CEO Richard Petrie visited the China BIM Union to discuss openBIM and P-BIM.

In June 2018, the discussion of the P-BIM theoretical system with Professor Richard Choy was held in Shenzhen, and the name of Breakdown and Coding System of AEC Industry Information finalized as A&bCode instead of formerly-proposed A&bClass.

I wouldlike to thank all the teachers and students of the China BIM Union Advanced Seminar.

The preface of this book was translated by Wu Lufang, the first chapter by Wang Zhaohua, the second by Duhao and Wu Lufang, the third by Zhang Miao, Wang Zhaohua, Xu Xiaoxiao and Yang Songqiao, the fourth by Zhao Feng, the fifth by Pu Qiuxing, the acknowledgements by Wang Zhaohua. The book was proofread by Wu Lufang, Wang Zhaohua and Pu Qiuxing. The book was compiled by Wu Lufang.

Thanks to all illustrators of this article: Zhang Liang, Sun Wencheng, Liu Julong, Zhou Bingsong, Zheng Jinzhou, Cai Yaocong.

Huang Qiang  
Jan. 19th, 2019

# Introduction of Advanced Training on BIM Application & Industry Collaborative Innovation

## 【 BACKGROUND 】

As a public welfare organization, China BIM Union (hereinafter referred to as the Union) composed of 16 union members was founded in January 2012. The 1st Advanced Training on BIM Application & Industry Collaborative Innovation by China BIM Union was hosted in September 2015. The training is held every six months since the beginning, and seven sessions were successfully completed.

Figure 1　Opening ceremony of the Advanced Training on BIM

## 【BIMER'S DNA】

Each participant of the Advanced Training on BIM should have 8 basic quality as follow:

- Challenge Taker
- Logical Thinker
- Science Lover
- Curious Person
- Problem Solver
- Big Thinker
- Pragmatist
- Societal Contributor

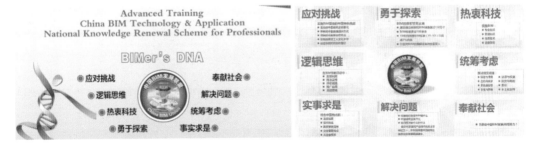

Figure 2　BIMER'S DNA

## 【 COURSE CONTENTS 】

The training lasts for 6 days, and the course contents includes 5 research topics:
1. Overview of BIM application and research progress
2. BIM technology research and achievement display
3. BIM applications development and project practice
4. Collaborative innovation platform of digital BIM industry
5. Completion result report

Figure 3　Key course content and 5 research topics

The main programs of the course, included keynote speeches, thematic studies, academic interflows, discussions with experts and site tours with full online live broadcasting.

Figure 4　Ice-breaking activities at the beginning of term

Figure 5　Huangqiang, Chairman of the Board of CBU, give keynote speeches

Figure 6　Academic interflows

## 【 FREE TO ATTEND 】

The fees of attend the advanced training courses, get all training materials and all-day meals are free of charge for each admitted students.

The expenses of the trainees include meal, training site and lecture expenses, and training materials shall be subsidized by the host organization.

Students only need to bear the cost of accommodation and transportation.

## 【 TRAINING RESULTS 】

The seven sessions advanced traininghas trained 560 BIM professional technical personnel and senior management personnel.It provides important talent support for the sustainable innovation and development of the AEC/FM industry, and plays a positive demonstration effect in leading the creation of a new pattern of talent education and training in china.

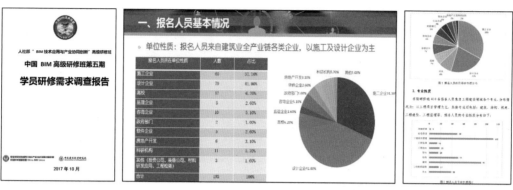

Figure 7　Online registration statistics

Figure 8　Team completion report

Figure 9　Honorary certificate of primier trainee

# 【APPLICATION PROCEDURE】

Overseas students are welcome to join the advanced training on BIM and become a member of our team.

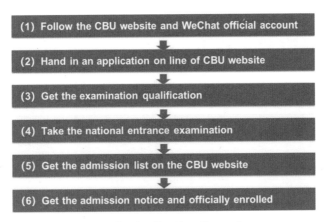

Figure 10   Application Procedure

Figure 11   Remarks by BIMer's

# 【 INVITATION 】

We sincerely invite experts from overseas industries, universities and research institutes to participate in the advanced training on BIM. Your attention and support are a great motivation for our efforts. We are looking forward to your coming.

Figure 12　Honorary instructors

Figure 13　The graduation ceremony of the Advanced Training on BIM

# 【CONTACT】

Figure 14    WeChat official account of China BIM Union

China BIM Union Secretary Office
Room 301 of Building B (B301)
China Academy of Building Research
No.30, Beisanhuan East Road
Beijing 100013
www.chinabimunion.org
chinabimunion@126.com
TEL: 010-64694968

# A&bCode Demonstration of Software Data Exchange

China BIM Union and Shenzhen University jointly established a data exchange laboratory (Internet of AEC Industry and BIM Lab) for P-BIM application. In the BIM Lab, the different software, in compliance with the standards for P-BIM software and information exchange, accomplishes professional design construction tasks and data interoperation through the collaborative management platform.

Dr. Calvin Kam, Invited Executive Directors ofChina BIM Union ,Program director and adjunct professor of center for integrated facilities engineering of Stanford university, and the Founder and CEO ofbimSCORE Limited, made the detailed explanation to the software and information exchange ,and data interoperation in the BIM lab platform.

## 【VIDEO】

Video download address on Baidu cloud (17 minutes):
https://pan.baidu.com/s/1kvNzt62wCAfViNl9G4C0aA
Code: 8r2h

Software and Standards Used in the Internet of AEC Industry and BIM Lab  Table 1

| No. | Profession | Software | Software developer | Standard |
|---|---|---|---|---|
| 1 | Project management | Project management cloud platform | Beijing Xunlian Ruichi Tech. Co., Ltd. | Standard for P-BIM software function and information exchange of EPC project management (under development) |
| 2 | Geotechnical investigation | LIZHENG | Beijing Leading Software Co., Ltd. | Standard for P-BIM software function and information exchange of geotechnical investigation T/CECS-CBIMU3-2017 |
| 3 | Foundation excavation design | LIZHENG | Beijing Leading Software Co., Ltd. | Standard for P-BIM software function and information exchange of foundation excavation design T/CECS-CBIMU4-2017 |
| 4 | Foundation design | PKPM | CABR Technology Co., Ltd. | Standard for P-BIM software function and information exchange for design of soil and building foundation T/CECS-CBIMU5-2017 |
| 5 | Architectural design | GRAPHISOFT | GRAPHISOFT | Standard for P-BIM software function and information exchange of architectural design (under development) |
| 6 | Concrete structure design | PKPM | CABR Technology Co., Ltd. | Standard for P-BIM software function and information exchange of concrete structure design T/CECS-CBIMU7-2017 |
| 7 | HVAC design | HONGYE | Beijing Hongye Technology Co., Ltd. | Standard for P-BIM software function and information exchange of heating ventilation and air conditioning design T/CECS-CBIMU11-2017 |
| 8 | Water supply and drainage design | HONGYE | Beijing Hongye Tongxing Technology Co., Ltd. | Standard for P-BIM software function and information exchange of water supply and drainage design T/CECS-CBIMU10-2017 |
| 9 | Electrical design | HONGYE | Beijing Hongye Tongxing Technology Co., Ltd. | Standard for P-BIM software function and information exchange of electrical design T/CECS-CBIMU12-2017 |
| 10 | MEP collaboration design | HONGYE | Beijing Hongye Tongxing Technology Co., Ltd. | Project approval |
| 11 | Concrete structure contract | TEKLA | Trimble | Project approval |

continued Table

| No. | Profession | Software | Software developer | Standard |
|---|---|---|---|---|
| 12 | Reinforcement contract | Reinforcement detail design | Xingceng Building Technology | Project approval |
| 13 | Formwork contract | PMS Formwork | Hangzhou Pinming Software Co., Ltd. | Project approval |
| 14 | Cast-in-place piles contract | Cast-in-place piles contract software | CABR Foundation Engineering Co., Ltd. | Project approval |
| 15 | Cast-in-place piles construction | Cast-in-place piles construction software | Shanghai Foundation Engineering Group Co., Ltd. | Project approval |

## 附图：A&bCode 与建筑业互联网

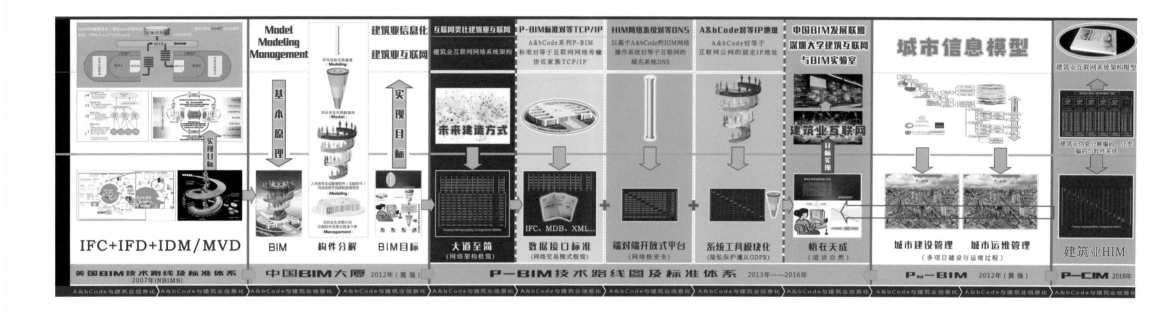

扫码看图